Praise for *Cows Save the Planet*

"Judith Schwartz's book gives us not just hope but also a sense that we humans—serial destroyers that we are—can actually turn the climate crisis around. This amazing book, wide reaching in its research, offers nothing less than solutions for healing the planet."

—Gretel Ehrlich,
from the Foreword

"Judith Schwartz takes a fascinating look at the world right beneath our feet. *Cows Save the Planet* is a surprising, informative, and ultimately hopeful book."

—Elizabeth Kolbert, author of
Field Notes from a Catastrophe: Man, Nature, and Climate Change

"In *Cows Save the Planet,* Judith Schwartz takes us on a fascinating, John McPhee–style journey into the world of soil rehabilitation. The eclectic group of farmers, ranchers, researchers, and environmentalists she visits have one thing in common: they all believe in the importance of organic matter in the soil for solving our most pressing environmental issues. Some of the innovative techniques they use to increase the vitality of their soil include no-tillage, using deep-rooted perennial grasses, cover crops, mulching, and, surprisingly, grazing large herds of animals according to a program called 'holistic management.' Imagine, a book about soil that's a real page turner!"

—Larry Korn, editor of *The One-Straw Revolution* and
Sowing Seeds in the Desert by Masanobu Fukuoka

"Here's a secret climate-change activists and energy-efficiency and renewable-energy promoters neglect: nature is designed to be self-healing, and her most profound 'tool' is photosynthesis. 'Free' sunlight is the best energy source to extract carbon dioxide from the atmosphere while also producing organic matter and oxygen—and a by-product is healthier soil, forests, wetlands, and ecosystems. When politicians, policy leaders, and activists get serious about cost-effective solutions to climate change, then a top priority will be ecological restoration to harvest and store carbon naturally, and Judith Schwartz's new book will provide a destination and map."

—WILL RAAP, founder,
Gardener's Supply and Intervale Center

"Judith Schwartz reminds us that sustainable range management is as much about the microbes in the soil and their feedback loops with cattle as it is about the cattle themselves. When I finally go home on the range to be composted, I want to be part of the miraculous cycle of rangeland renewal that is managed in the way that Schwartz describes so well."

—GARY PAUL NABHAN, author of *Desert Terroir*,
Kellogg Endowed Chair in Sustainable
Food Systems, University of Arizona

COWS SAVE THE PLANET

AND OTHER IMPROBABLE WAYS OF RESTORING SOIL TO HEAL THE EARTH

JUDITH D. SCHWARTZ

FOREWORD BY GRETEL EHRLICH

Chelsea Green Publishing
White River Junction, Vermont

Project Manager: Bill Bokermann
Developmental Editor: Brianne Goodspeed
Copy Editor: Laura Jorstad
Proofreader: Helen Walden
Indexer: Margaret Holloway
Designer: Melissa Jacobson

Printed in the United States of America.
First printing April, 2013.
10 9 8 7 6 5 4 3 14 15 16

Our Commitment to Green Publishing
Chelsea Green sees publishing as a tool for cultural change and ecological stewardship. We strive to align our book manufacturing practices with our editorial mission and to reduce the impact of our business enterprise in the environment. We print our books and catalogs on chlorine-free recycled paper, using vegetable-based inks whenever possible. This book may cost slightly more because we use recycled paper, and we hope you'll agree that it's worth it. Chelsea Green is a member of the Green Press Initiative (www.greenpressinitiative.org), a nonprofit coalition of publishers, manufacturers, and authors working to protect the world's endangered forests and conserve natural resources. *Cows Save the Planet* was printed on Natures Natural, an FSC®-certified, 30-percent postconsumer recycled paper supplied by Thomson-Shore.

Library of Congress Cataloging-in-Publication Data
Schwartz, Judith D.
Cows save the planet and other improbable ways of restoring soil to heal the earth / [by Judith D. Schwartz].
p. cm.
Includes bibliographical references and index.
ISBN 978-1-60358-432-6 (pbk.) — ISBN 978-1-60358-433-3 (ebook)
1. Soil ecology. 2. Soil restoration. I. Title.

QH541.5.S6S425 2013
577.5'7—dc23

2013001703

Chelsea Green Publishing
85 North Main Street, Suite 120
White River Junction, VT 05001
(802) 295-6300
www.chelseagreen.com

To Tony and Brendan

Contents

Foreword
Going to Ground

We "go to ground" when exhausted by disaster or war, when we need to restore ourselves, look natural beauty in the face, and nourish ourselves by growing food; we go to ground to seek solace. Now, as we find ourselves facing a grave threat to civilization—the global emergency of climate change, desertification, and habitat destruction—we would be wise to go to ground to find how we might survive.

Ground in this sense represents not only basic sanity, but actual soil and all the life-giving processes that emanate from it. Nature is matrix and embrace. Photosynthesis is foundational, our only true wealth. Without it, we devolve. Poor land leads to poverty, hunger, social unrest, cultural deprivation, inhumanity, and war. So we must wonder why the biological health of the planet is not our number-one priority. In our careless, destructive, and proprietary ways, we have ignored the biological requirements of the living planet, and as a result of our neglect and abuse ground has become, alternately, a hot plate, a desert, a crumbling sea cliff, and a floodplain.

Judith Schwartz's book gives us not just hope but also a sense that we humans—serial destroyers that we are—can actually turn the climate crisis around. This amazing book, wide reaching in its research, offers nothing less than solutions for healing the planet.

Almost thirty years ago I was asked by *Time* magazine to write about visionary thinkers in the American West. One of those I chose was wildlife biologist, game rancher, and restoration ecologist Allan Savory, now in his midseventies and founder of the Savory Institute, who figures prominently throughout this book.

Savory, at age twenty, was put in charge of wildlife in a large part of Northern Rhodesia (now Zambia). In those years he began puzzling over the root causes of habitat destruction, and the needs of wildlife, domestic livestock, and humans living together on the land.

To have followed Savory from his days in the bush, through the horrendous civil war during which he commanded a tracker combat unit and led the opposition against the racist government of Ian Smith, and on to his eventual emigration to the United States, is to have watched a man thinking and rethinking through the problems of how to heal the earth.

Now the recipient of many awards and millions of dollars in funding from international sources, to help put things right in places like Kenya, Australia, the United States, Mexico, and South Africa, Savory provided the initial "kick in the ass" for many younger many ranchers, farmers, ecologists, and scientists. The people Schwartz interviewed for this book aren't theorists; they practice what they preach.

Since I met Savory in the 1980s, the health of the planet has deteriorated seriously. Too few paid attention or took action. We now have a global emergency on our hands: climate change and the desertification of the earth's surface.

Savannas are drying, Arctic coastlines are being eroded by retreating ice and stormy seas, dry northern valleys are being pummeled with unseasonal rain followed by drought. Tundra around the top of the world is melting; rain forests are drying; the great Australian drought is spreading to its verdant edges. Tree mortality, especially from Mexico to the Yukon, is rampant, and aquifers are being drained.

The jet stream has been destabilized, and weather systems have become chaotic. Deep winter cold or searing heat sticks in one place for prolonged periods, with no clearing winds sweeping it away. As a result, storms pound down and cause unthinkable destruction.

We've been watching the shocking rise of greenhouse gases in our atmosphere: CO_2 and methane from smokestacks and tailpipes, from thawing permafrost, and from thermal heating of the oceans that causes methane clathrates to rise in plumes straight out of the East Siberian Sea. We've experienced the devastation of violent storms: hurricanes, tornadoes, and typhoons, as well as wildfires and floods. Yet we fail to make sense of it, because too few of us have an intimate relationship with the natural world—as if we were something other than "nature."

We can deny climate change because to do otherwise would imply that we have to tear the global economy apart, which no one can do; we can persist in thinking that creeping deserts and melting ice have nothing to do with us because we have failed to think globally and holistically, or understand that the Arctic drives the climate of the world. We can put our heads in the sand because it's painful to hear that we have enabled a failing civilization. We can bemoan the ever-increasing parts per million of CO_2 in the atmosphere, yet fail to look at the whole carbon picture. We've already stumbled over the tipping point. All we can do now is deal with the consequences.

Go back to the word *root*, and you'll see what we've been missing: the soil under our feet, and how it functions in the climate. We've been obsessed with fossil-fuel emissions, but we have not taken into consideration the ground we stand on; what's been taking place there; or the carbon, solar, water, and mineral cycles in the soils of savannas and grasslands. Time to go to ground.

You might ask what dirt has to do with global warming. In reading this astounding book, we will learn how to unmake deserts, rethink the causes of climate chaos, bring back biodiversity, and restore nutrients to our food. In other words, how to stanch and heal the great wound we have inflicted on our planet.

It's possible, and it starts right under our feet.

Every idea and solution in this book is nature-based. Going to ground, in this context, involves understanding the anatomy of a piece of ground, its microbial makeup, and the way soil and plants can eat sunlight and carbon, or hold water. Join me in gaining a radical understanding of the root causes of desertification, species extinctions, and global heating. In doing so we can begin to understand the restorative cycles of carbon, water, minerals, and sun that will help us heal our planet.

Widen your mind with a holistic approach to the extinction cliff. No advanced technology necessary: only common sense, intuition, hard work, and a desire to try the new. Don't stop here: Dive into the improbable.

Learn about "good carbon" from Australian soil ecologist Christine Jones, who writes: "Carbon is the currency for most transactions within and between living things. Nowhere is this more evident than in the soil." Learn how good carbon held in topsoil and therefore grasslands—which cover 45 percent of land worldwide—increases biodiversity. That fertile soil and its rich microbial life hold water, thus restoring the water table. This is how good carbon gives us life, Jones writes.

We meet farmers who direct-drill oats into native grasslands with chisel-shaped Keyline plows instead of stripping the ground, then plowing and replanting with an annual crop. We learn that topsoil is America's largest export. Wind carries dirt from plowed farm fields thousands of miles away—and with it all the carbon, moisture, and nutrients that belong under the ground.

Interviewed in Schwartz's book are ecologists who show that it is possible to actually *make* topsoil with good farming practices, instead of waiting thousands of years for geological weathering to take place.

Learn from Cornell scientist David Pimentel, according to whom "Ninety percent of our cropland is losing soil to wind and water erosion at thirteen times the rate that soil is being formed." Then read about Australian rancher Colin Seis, whose farm Winona has been made famous for its "accelerated soil creation." Learn about the "lower depths" where, deep down, humus is created and retains minerals and water.

Learn about "liquid carbon pathways" that help deliver the sun into the earth; understand that photosynthesis is our only true wealth, because without it, the soil becomes so degraded that the food we grow lacks nutrients, rain runs off, drought occurs, deserts grow, and inevitably people are pushed away.

Learn that by managing land with water circulation in mind—water, water vapor, condensation, and the entire water cycle from atmosphere to ground, up through the plant and back into the atmosphere, all of which the author calls "the continual back-and-forth drift of moisture between land and sea"—we will achieve nothing less than ending desertification, recharging aquifers, and restoring balance to the destabilized elements in the climate: carbon and water, and therefore the jet stream and the umbrella of greenhouse gases that holds heat in. Reversing climate change may have less to do with CO_2 than with raising H_2O in the atmosphere.

In these pages we see the interplay between the large water cycle and the small, the biotic pump, the flux of evaporation and condensation in the air and in plants and soil, and the constant exchanges between the two. We learn to think of soil as "a huge basin for water." It's important to think of restoring the climate drop by drop and managing land with water foremost in mind, because soon demand for water will be 40 percent greater than what is available.

If grasslands cover more than a third of the Earth's surface, and herbivores co-evolved with native grasses, then we must put animals back on the land: yaks, bison, caribou, elk, deer, kangaroos, antelope, pigs, horses, sheep, and cattle. "Land needs the presence of animals," Allan Savory says. Domestic livestock is managed to mimic wild herbivores. Overgrazing is a function of time, not numbers of animals. You can put a thousand head of cattle in a fifty-acre pasture for six hours in the spring and grass will thrive. But put one or two cows on a thousand acres for six months, and you'll overgraze the whole thing.

Like organic gardeners, herbivores aerate, nourish, and graze the land in ways that regenerate all the basic building blocks, increase biological activity, and increase productivity. Management of time, frequency, and intensity of grazing, plus a less stressful way of handling animals, serves to stop desertification. In this way, health is restored to million of acres of degraded, starved, and abused land.

Did I say this was going to be status quo thinking? Dive in and read on. The preceding is just a taste of what's in this wonderful volume. By learning to regard whole ecosystems and the wholes within wholes, we can make better decisions about how to manage land and animals.

By the end of Schwartz's book, we come to understand in real terms the links among the health of economic, societal, and ecological cycles: that one cannot be healed without healing all the others. We learn to look at the whole with all its components, while solving each particular within the whole.

After taking Savory's weeklong seminar in the early 1980s, I changed the way I ranched and handled cattle, and the way I thought about grass, sunlight, water, and soil. I began considering the whole ecosystem, not just bits and pieces, each of which needed help; I made a land plan, moved cattle every three days through smaller pastures divided by portable electric fence, and in two years restored twelve to fifteen species of native grasses to the range, developed springs, provided grass that sustained a winter herd of elk, stopped all overgrazing, restored watersheds, improved cow and calf health, maintained against all odds a 100 percent calving rate, and grew a large garden from which we ate year-round. Because I wasn't born to a ranching family, I could adopt new ideas without resistance. I planned, monitored, and replanned. I made mistakes, saw what went wrong, and corrected them. I was free to be as creative and collaborative in my problem solving as possible. It made ranching fun.

Judith Schwartz has picked up the many threads of new thinking that have developed since then and explored them with grit and eloquence. Her probing questions go straight to the point: How do we reverse global warming and habitat destruction? How do we create a healthy environment that benefits all sentient beings? How do we reverse our devolution? How do we survive? And if you think you already know all about it, I encourage you to start fresh: Judith's book spills over with a whole new generation's ideas.

I hope you will join me and be courageous in exposing yourself to new thinking, new knowledge. The improbable will smooth out into the possible, the actual. Read it once, then again, then try it out in your own backyard as I did. Why not have solar panels on every building, radiant heat, and water catchment systems? Why not vegetable gardens instead of asphalt rooftops or unused lawns?

The earth is a living membrane, a fragile skin deeply responsive to our every action and footstep. We breathe in weather; we breathe out CO_2. The destruction and disappearance of Arctic and Antarctic ice is not a fairy tale happening in some far-off place. These are whole ecosystems in a state of collapse. Every whole is connected to every other whole. Fresh water from melting ice sheets pours into the whole-ocean circulation system. It alters salinity and disrupts the Gulf Stream; it raises the amount of water vapor in the air and the oceans, which in turn alters weather and the entire climate. The albedo effect of ice in the Arctic makes it the natural air conditioner of the earth. Without it, crops, water, and sentient beings won't survive.

Lately I've been studying a bit of classical Chinese for the ways their words and thoughts derive from the natural world and help us understand its workings. One set of characters means, simultaneously, sun and moon, to be intelligent, to understand, and tomorrow. Another one means: breath-seed-life.

Tuck those in your pocket and go roaming. See how you can regenerate sunlight, carbon, water, and minerals into and out of the soil. Test your ideas against the basic building blocks that keep land and animals healthy. Do not try to save just one species. Save the ground under their feet first, and in doing so you'll bring the whole environment to health.

Exchange yourself for others: for the savanna, a blade of grass, a pride of lions, or a vole; or for a grazing cow or a caribou. Then your suit of armor will soften, and resistance to new ideas will fall away. You will come on fresh ground. You won't have to pursue happiness; happiness will pursue you. Go to ground, embrace the whole, come alive.

GRETEL EHRLICH
FEBRUARY 2013

Introduction

> There can be no life without soil and no soil without life: they have evolved together.
>
> —Charles E. Kellogg, *Soil and Society*, 1938

COWS SAVING THE PLANET? Why not? An idea that sounds preposterous begins to make sense if we stop to take a soil's-eye view of our current environmental predicament. To crouch down to ground level—literally or metaphorically—and see how human and animal activity enhances or does violence to that fine earthy layer that hugs our planet. To appreciate the imperceptible animal–vegetable–mineral dance that keeps us alive.

You see, that brown stuff we rush to wash off our hands (or, depending on our age, our knees) is the crux of most biological functions that sustain life. Soil is where food is created and where waste decays. It absorbs and holds water; or, if exhausted of organic matter, streams it away. It filters biological toxins and can store enough carbon to reduce carbon dioxide levels significantly and relatively quickly. It is home to more than 95 percent of all forms of terrestrial life. In any given place the quality of the soil greatly determines the nutritional value of food, how an area weathers drought or storms, and whether an ecosystem is teeming with life or the equivalent of a ghost town.

Where do those cows fit in? Cattle, like all grazing creatures, can, if appropriately managed, help build soil. When moved in large herds according to a planned schedule, livestock will nibble plants just enough to stimulate plant and root growth, trample the ground in a way that breaks apart caked earth to allow dormant seeds to germinate and water to seep in, and leave dung and urine to fertilize the soil with organic matter (aka carbon). The result is a wide variety of grasses and other deep-rooted plants and rich, aerated soil that acts like a great big sponge so as to minimize runoff and erosion. (Cows and their eruptive digestion habits have gotten a bad rap of late—I'll address the methane question in chapter 1.) The use of ungulates such as cattle in land restoration, a practice called Holistic Management, was developed and refined over the decades by Allan Savory, a farmer and rancher and former opposition leader to then-Rhodesia's white government. With cows or other grazers operating under Holistic Management across large areas of degrading land, this could mean a great deal of soil created or preserved.

Leaving behind our bovine herd for the moment, another way to build soil is through zai pits, a traditional growing method from Burkino Faso in West Africa. Small holes are dug into a field, and these capture water and hold soil organic matter (compost and such), both precious resources in drylands that depend on seasonal rainfall—about a third of the world's landmass. Cattle have a similar impact. Rancher and consultant Jim Howell told me that this helped Grasslands, LLC's, South Dakota ranches withstand the spring 2011 torrential rains while nearby properties suffered losses: The herds left hoof-size pockets in the ground, so water pooled rather than forming gullies and eroding the land.

If you're wondering why we want to build soil—isn't there enough dirt out there already?—consider this: Around the globe, we're losing topsoil somewhere between ten times (in the United States) and forty times (China and India) faster than we're generating it, some eighty-three billion tons of it a year. Soil is pounded off fields during a rainstorm; it runs down our rivers; its surfaces are over- and undergrazed; when left uncovered it loses its organic matter as carbon oxidizes and enters the atmosphere. Despite our collective societal indifference to soil, we've all got a large stake in its fortunes. In an oft-quoted and paraphrased line, "Man has only a thin layer of soil between himself and starvation." Up to now, we've been heedless with our soils. And we're paying the price.

On an immediate, day-to-day level, the food we eat is only as good as the soil from which it springs. In part because of soil depletion, most food grown today is less nutritious than that of most previous eras. Research from the UK Ministry of Health determined that a steak today has half the iron of its counterpart fifty years ago thanks to changes in what the animals eat. Breeding crops for high yields accelerates the dilution of nutritional content. Over time this can lead to nutrient deficiencies, which a grower may not notice until the effects on the plants are visible, by which point the situation has become extreme.

Remember the adage "An apple a day keeps the doctor away"? Over the last eighty years, the calcium content of one medium apple has dropped by nearly half, and levels of phosphorus, iron, and magnesium have fallen more than 80 percent. So to get the same doctor-avoiding kick, you'd now need four or five apples. And this is fruit straight from the tree; processed foods also lose nutrients en route from the field to box or bottle. Some scientists believe today's high obesity rates are, paradoxically, a symptom of malnutrition due to diets deficient in micronutrients. Which prompts the question: Could the declining nutritional content of our food also be a factor in our rising rates of chronic diseases and allergies, particularly food allergies among children?

Fortunately, a host of creatures underfoot are ready to make and enhance soil for us—once conditions are right. This is where that microscopic choreography comes in; the cows (or the diggers of holes) are only the catalyst. Worms, insects, and microorganisms like fungi and bacteria aerate the ground, decompose waste, exchange nourishment (mycorrhizal fungi take glucose from plants and in return help plants assimilate nutrients), and break down rocks into minerals like calcium, magnesium, iron, and zinc that are essential to our health. The herbicides, pesticides, and fungicides widely used in industrial agriculture kill many of these organisms; from the soil's or soil dweller's perspective, chemical additives are not such a great thing.

With zai, the organic matter in the hollows attracts termites. The termites, in turn, burrow around and create tunnels, allowing water to penetrate the ground rather than evaporate. Though usually regarded as pests, termites in marginal lands play much the same role that earthworms do in greener climes.

In *Dirt: The Erosion of Civilizations*, geomorphologist David Montgomery offers numerous cautionary tales of kingdoms, cultures, and empires that squandered their soil and found themselves with nothing left to live on. From the earliest farmers in the Fertile Crescent to the Mayans, Romans, and Easter Islanders, societies have exhausted their land either to scatter and regroup in much-diminished form, or to become lost to history.

Not that people didn't know better. Advice about caring for soil has been passed along since the first primitive hoes broke virgin ground.

Luc Gnacadja, executive secretary of the United Nations Convention to Combat Desertification, likes to quote this proverb from the Sanskrit Vedic Scriptures of around 1500 BCE: "Upon this handful of soil our survival depends. Husband it and it will grow our food, our fuel and our shelter and surround us with beauty. Abuse it and soil will collapse and die, taking humanity with it." More recently, in 1937 Franklin D. Roosevelt made the same point with a nationalistic twist: "A nation that destroys its soils destroys itself."

Despite history's warnings, the temptation to plant on fragile hillsides, clear forests, push yields of lucrative crops, or otherwise try to squeeze more from the earth proves too great. But today we can't just pack up our tent and move to more promising turf while leaving the damage behind us.

It's time to start treating soil as the precious resource it is. This doesn't mean forgoing its bounty—soil is a renewable resource that can respond quickly to watchful stewardship. Since soil is integral to so many biological processes, nurturing and improving it provides us with many paths toward ecological renewal—with returns far greater than what you'll see at your feet.

Let me clarify right up front: If a few years ago someone told me I'd be writing about soil, much less be fascinated and excited by it, I'd have said they were crazy. Wait, I'll correct that—since I'd begun reporting on the New Economics and the juncture of economics and the environment, I'd grown accustomed to finding my thinking stretched in new and improbable directions. As a suburban-raised writer who can barely tend a houseplant, I'd rank as among the least likely of guides to this rich, brown mantle of our planet and its potential role in restoring our environment. But bear with me as I make the case that soil can be seen as the crucible for our many overlapping environmental, economic, and social crises (excess atmospheric carbon dioxide, drought, floods, wildfires, food scarcity, desertification, and obesity/malnutrition). And that focusing on soil restoration will allow us to begin to chip away at these seemingly insurmountable problems.

Professional pursuits are often driven by personal, emotional needs. For me, over the last few years, that has been a need to allay anxiety, specifically anxiety about the environment. I'd think about the wildlife I'd taken for granted and lament the diminished natural wealth that future generations would inherit. Every day's news brought upsetting developments—ice caps melting, northeastern bats dying of white-nose syndrome, bee colony collapses—and I realized I had two choices: erect a mental blockade and ignore it all, or find a way to engage with it. I chose to engage as a journalist. I was determined to find solutions, ideas that sparked optimism so that I wasn't tempted to barricade myself.

For a while I focused on environmental economics. Slow Money, which encourages investment in local food enterprises, introduced me to an appreciation of soil as wealth. I started writing articles on soil restoration—and kept bumping into the kind of encouraging ideas I was looking for but wasn't seeing in the media. In the course of reporting on Holistic Management, I began to realize that for every seemingly unsolvable problem there was a flip side, an alternative set of strategies that would restore balance to the system. And that the question of which way it went often turned on soil.

For example, our carbon problem. When you hear reports of rising carbon dioxide levels, it's easy to get the impression that the carbon-and-oxygen molecule is a kind of toxin, some alien vapor coughed up by a century-plus of rampant industrialism that has now come back to haunt us. But the trouble isn't the carbon itself; it's that there's too much of it in the air rather than in the ground, where it lends fertility to the soil. Soil, it turns out, is the natural and the most cost-effective carbon sink. According to Rattan Lal, Distinguished University Professor at the Ohio State University, soil carbon restoration can potentially store about one billion tons of atmospheric carbon per year. This would offset around 8 to 10 percent of total annual carbon dioxide emissions and one-third of annual enrichment of atmospheric carbon that would otherwise be left in the air.

Consider also biodiversity, which starts in the soil; there are as many living organisms in a teaspoon of healthy soil as there are people on the planet. The multiplicity of life underground is reflected in what grows and flourishes on the land. Conversely, deficiencies in the soil limit the plant and animal species that can survive above, from the simplest forms and on up the food chain. Once identified, such imbalances can be remedied, restoring the full range of life to that particular place.

And let's look at water. Water has increasingly become a problem, with at once not enough in some parts of the world and way too much in others. Sometimes too much comes right after not enough so we get torrents of rain coursing down parched, barren land with nothing to slow it down. Healthy, living soil can contain many times its weight in water. Steven Apfelbaum, a restoration ecologist in Wisconsin, says that every 1 percent increase in soil carbon holds an additional sixty thousand gallons of water per acre. Not only does this limit damage from erosion, but it also keeps water on the land. This feeds plants, builds aquifers, and maintains the moisture that promotes microbial life.

There will always be floods and droughts and so-called hundred-year weather events that are now happening with unnerving regularity, but they needn't cause such devastation. I spoke to Zachary Jones, of the Twodot Land and Livestock Company near Harlowton, Montana, where in spring 2011 the Musselshell River, a tributary to the Missouri River, saw thirteen times its usual spring runoff: "The most water we've ever seen in that creek in the five generations my family has ranched here." While the flood closed highways, washed away barns and corrals, and drowned livestock, Twodot's land remained unscathed, with very little runoff. Jones attributes this to its having been under Holistic Management for twenty-five years. Compared with its neighbors, Twodot's twenty-four thousand acres had a greater variety of grasses and other plants with deep roots (for efficient nutrient and water cycling) and rich, aerated, highly absorbent soil. Rather than streaming off and causing erosion, the water stayed on the land.

Here was cause for hope: Focusing on soil could minimize the devastation of floods. I couldn't get enough of that optimism, so I kept exploring. It turns out that accelerated topsoil formation is doable, with the tools required as modest as livestock, hole digging, organic amendments, and simple cultivation techniques; no costly, high-tech, geo-engineered you-need-a-PhD-to-understand-it schemes. I began to see that, despite the way our ecological predicaments are often portrayed, we aren't dealing with discrete problems to be tackled one at a time. Rather, our environmental messes are symptoms of disrupted biological cycles: the carbon cycle, water cycle, nutrient cycle, and energy cycle. Since all these cycles are interconnected, efforts to redress one system will likely help restore the others.

I don't mean to come across as naive, or to suggest that we can throw some cattle and compost on the ground and go on wasting and polluting as before. But neither am I willing to be paralyzed by despair, nor take refuge behind that barricade of indifference, no matter how tempting at times. I know how bad things are. But we've got to start somewhere. Soil restoration can be done anywhere: one watershed, one community, one abandoned field. At whatever scale, attend to the needs of the soil, and the ecological cycles will begin to get back in sync.

One sweeping and dramatic example is the restoration of the Loess Plateau in China, documented in John D. Liu's film *Hope in a Changing Climate*. Over ten years, an area the size of Belgium along the Yellow River in northwest China was transformed from a near-barren desert plagued by dust storms, considered the most eroded place on earth, to a thriving agricultural region with the poverty rate lowered by half. Through a government–community partnership, local farmers built terraces, reforested sloping land, and shifted to perennial crops that have deeper roots. Basically, the Chinese government saw that it would cost less to stabilize the soil than to continually deal with the sediment running into the river. In stabilizing the soil, numerous other benefits followed.

In considering our environmental challenges, one wild card is the extent of nature's ability to heal itself. The awareness that restorative feedback mechanisms may be greater and more powerful than we appreciate gives me tremendous hope. A few years back I attended the annual E. F. Schumacher Lectures at what is now the Schumacher Center for a New Economics in Great Barrington, Massachusetts, not far from where I live. (Schumacher was the fellow who wrote the early-1970s book *Small Is Beautiful*, with the evocative but oft-ignored subtitle *Economics as if People Mattered*.) At the end of the day I was standing around among a group chatting with Alisa Gravitz, executive director of Green America, who'd given a talk called "Everyone Is an Activist." The speakers, including environmental stalwart Bill McKibben, had done a fabulous job of articulating just how dark the state of our environment was, and I think many of us were looking for some flicker of hope to carry home. Gravitz has an easygoing, upbeat style, so several of us were drawn to her.

One woman asked straight-out how to keep going in the face of constant bad news. Gravitz thought for a minute, then said that she grew up near Lake Erie, which in the 1960s was declared "dead." (Famously, in 1969 Cleveland's Cuyahoga River, a tributary to Lake Erie, burst into flames.) They were told that restoring the lake would take decades, Gravitz recalled, but very quickly signs of life appeared. With tightened pollution restrictions, including the Clean Water Act in 1972, the lake bounced back and by 1980 was a popular recreation destination. There are now large algal blooms and occasional fish die-offs. But Gravitz's point was that the lake most experts had given up on improved faster than anyone anticipated. "That's what I remember whenever I start to lose courage," she said. "We should never underestimate ecology's capacity to heal."

This made sense to me: We need to work with the natural inclination of all living things to strive toward health. We can start with one lake, one watershed, one devastated plateau, and create isles of reparation. Then, over time, fill in the gaps, jigsaw-puzzle-like. Every stretch of depleted or chemically saturated soil, every gulch and gully carved by erosion, is a wound on the landscape. We can heal those wounds by tending to the soil.

When I was a kid, we called it "dirt." It was brown (that unfortunate color with scatological overtones) and got stuck under our fingernails when we played outside, making the immediate, requisite hand washing an ordeal. It was an affront to the smooth-surfaced chrome aesthetic that dominated the era. There was the impression that if we failed to scrub it off, something bad would happen (if nothing else, our clothes might be "soiled"). Still, we played in it, with toy versions of the construction vehicles that were ripping up the earth just yards from our house. For this was the mid-1960s, and people were planting houses the way a thousand miles west in the Grain Belt they planted corn, acres and acres of homes sprouting up with their tidy brick chimneys and matte aluminum siding, consuming the land in architectural and social monocultures. We neighborhood kids were creatures of our time: In our pretend games we were builders, not farmers.

Even living in Vermont, where you're never far from a contented cow or, in the summer and fall, a farm stand (often a shed with a cash box, run on that old-fashioned principle, trust), it took me a long time to understand that soil is more than a one-size-fits-all passive medium for stuff that grows. A while back, I was chatting with our neighbor Charles Moses, who brush-hogs our meadow so that it stays meadow, and who teaches chemistry in our local high school and is a great-grandnephew of Grandma Moses. "You know," he said, nodding down toward our property, "you've got great soil down there."

The way he said it, I could tell that it was bugging him that we weren't growing anything on our land. I said, "Really?"

"Yup," he said, with the native Vermonter's habitual restraint. "If you'll put in some vegetables I'll till it for you. There's nothing like eating potatoes straight out of the ground."

There was no turning back. We bought vegetable starts at Clearbrook Farm, an organic gardening mecca in nearby Shaftsbury, and with scarcely any effort on our part things grew. A bit of beginner's luck; that hyper-fertile soil, lovingly nurtured by the farmer who lived there in the 1970s when Charles was growing up, had been waiting for some crops worthy of nourishing. I'm now an avid reader of seed catalogs. I so revel in the summer ritual of surveying the garden and harvesting that during those precious months, I'm reluctant to go away. On the agenda right now: a second planting of kale, beets, and beans, and what to do with all those cucumbers.

This book isn't a scientific or how-to guide to the soil. Rather, this is a call to action on behalf of the soil—and, by extension, those of us who benefit from it. Which is, of course, all of us. I'm not going to mince words or hide behind euphemism; we know ecological systems are crumbling all around us. Chapter by chapter, I'm going to hit our most severe environmental problems and explore how zeroing in on soil can move us toward solutions—and I'm going to share examples of people who are doing this through innovative, and often counterintuitive means. (Remember those cows?)

So roll up your sleeves and pant legs and wade with me into the dirt.

Chapter One

Ground Zero for Carbon Dioxide Reduction Is the *Ground*

> The process that actually removes CO_2 from atmospheric circulation is photosynthesis.
>
> —Christine Jones, Australian soil ecologist

WE'VE GOT A CARBON PROBLEM. Scientists tell us we've already passed the safety threshold for the concentration of carbon dioxide in our atmosphere. To avoid destabilizing the climate and provoking other associated catastrophes, we need to bring that down to 350 parts per million. Right now we're at 392 and creeping upward.

While we strive for "carbon neutrality" and scratch our heads over proposed carbon trading schemes, the problem isn't the amount of carbon per se—the quantity of carbon on earth is constant—but where that carbon is. The surplus, climate-stressing, troublemaking stuff is what enters the atmosphere when it combines with oxygen to form carbon dioxide gas.

Well, where are we supposed to put all that extra carbon?

Into the soil. Carbon is what lends fertility to soil and sustains plant and microbial life. Soil that's rich in carbon holds water, like a sponge. By contrast, water that falls on soil depleted of carbon streams off, causing erosion and leaching out nutrients. Land that retains water is more resilient to drought, wildfires, and flooding. In the words of Christine Jones, an Australian soil ecologist known Down Under as "the carbon queen," "Carbon is the currency for most transactions within and between living things. Nowhere is this more evident than in the soil."

Our crisis of excess atmospheric carbon dioxide is real and urgent.[1] It's a problem the likes of which we've never seen, at once invisible and global, abstract and yet far-reaching in consequence. But by focusing our efforts as we have been on the atmospheric component—trying to cap the lid on what we've been spewing into the air—we're only seeing half the picture. The other part of the story is that much of the legacy carbon hovering in the atmosphere is supposed to be down in the soil.

Putting Excess Carbon Where It Belongs

This knowledge sent me on a quest to learn more about "soil carbon"—a phrase that under ordinary circumstances would be a cue during a science lecture that it was time to drift off. And so I spent the better part of two days with two men devoted to the cause of soil carbon: Peter Donovan and Abe Collins, founding members of the Soil Carbon Coalition, a nonprofit whose mission is to "advance the practice, and spread awareness of the opportunity, of turning atmospheric carbon into soil organic matter." The tagline on the group's website reads: "Put the carbon back where it belongs."

Peter was passing through Vermont as part of the Soil Carbon Challenge, a competition to gauge how well land managers can facilitate the biochemical magic of drawing carbon back into the soil. The idea is to measure the baseline carbon levels on the properties of "contestants"—among them ranchers, farmers, and environmental nonprofits from California to Iowa to the Carolinas—and retest and document changes over the course of ten years. By focusing on small, fixed areas, the Soil Carbon Coalition organizers can monitor soil carbon changes accurately and with minimal cost using well-established methods of field sampling and laboratory testing.

How do you build carbon in the soil? By reversing the processes that released carbon into the air. Oil, coal, and gas represent one source of emissions, but over time the greater culprit has been agriculture. Since about 1850, twice as much atmospheric carbon dioxide has derived from farming practices as from the burning of fossil fuels (the roles crossed around 1970). In the past 150 years, between 50 and 80 percent of organic carbon in the topsoil has gone airborne. The antidote to this rapid oxidation is regenerative agriculture: working the land with the goal of building topsoil, encouraging the growth of deep-rooted plants, and increasing biodiversity. This turns the conventional approach to farming upside down: Rather than focusing on growing crops, the intention is to grow the soil. But "carbon farmers" like Donovan and Collins contend that as you build carbon levels, the rest—land productivity, plant diversity and resilience amid changing conditions—will follow.

Peter, fifty-nine, a soft-spoken former sheep farmer and scholar of classics and music, took the challenge on the road in July 2011. He gave up his apartment in Enterprise, Oregon, along with the bulk of his worldly goods and set out in a refitted 1981 school bus to monitor the nation's soils. He had several scheduled monitoring stops as he trekked east, but left time open to give talks and workshops and recruit more land managers and landowners to the competition. The monitoring and lab work cost $250, and, as he admits, offer no direct financial gain to participants—only the chance to be part of an effort to improve the land and environment.

One day in late October I drive up to St. Albans, Vermont, where Abe Collins is managing about 185 acres, most of which is used for cattle grazing. The air is raw and chill; the gray-on-gray sky hints at the freak early snowfall that will hit Vermont later in the week. The hills all around are dominated by the "burnt" colors (burnt umber, burnt sienna) that I remember ruled the rarefied autumnal corner of the crayon box. I wait in front of a farmhouse by the big yellow bus, with Oregon plates and a sign that reads WHOLE GRASSLANDS BUILD SOIL, as well as an offer to tune pianos for a reasonable fee, until a John Deere Gator rattles up. Abe, a younger man in jeans, jumps out and welcomes me. He wears a green-rimmed baseball cap with the words PLAYS IN THE DIRT. He looks down at my feet and frowns. "It'll be pretty wet out there," he says, and goes inside his house to find some women's boots he has around, one pair of which fits well enough. The fields, he says, are still saturated after the flooding from Hurricane Irene, almost two months before.

We ride the ATV toward Lake Champlain, a cool blue marker just over a mile away, and stop at a back pasture. I watch the two set up a transect and lay out a four-by-four-meter (thirteen-by-thirteen-foot) plot. They put down a metal hoop to create a kind of spatial "snapshot" that will allow them to zero in on a microcosm of the field. Peter and Abe kneel down and take a read of the land.

"There's dandelion, Italian ryegrass, blue grass—Kentucky and Canadian," Abe reports. "Meadow fescue. Clover, red and white. Very small amount of reed canary." All I can see is a bunch of grass with the occasional dandelion leaf bouquet.

The diversity of plants, Abe tells me, indicates the improved condition of the land. "When we started on the land, about six years ago, it was pretty badly drained and swampy." At the time, he explains, it was dominated by reed canary grass, an invasive species that has been driving out native plants in many parts of the United States, including Vermont. "I started loosening soil with a subsoiler, then grazed it and grazed again. Took hay crops, spread compost, laser-leveled it with a large plane, used a Keyline plow to aerate it. Before, the topsoil was four to five inches thick. At that point you got to an orange zone, the bottom of the topsoil where it's alternately aerobic [with oxygen] and anaerobic [without it]."

Abe shows me what soil at the oxygen threshold looks like: mucky wet clay with seams of rusty orange running through.

He continues: "Now there's eighteen inches of topsoil. We changed the conditions so there's oxygen and looser soil." And a greater variety of grasses and other plants, which reflects the land's ecological health and resilience.

Even more important than plant diversity is the fact that there are plants there at all, as opposed to bare ground. In soil carbon terms, that's a no-no. As Peter says, "Bare, uncovered soil indicates that there's not only leakage of soil carbon into the atmosphere, but the absence of life that can replenish it." Think of the huge tracts of agricultural land after corn or soybeans have been gathered up, all that naked dirt lying there as carbon wafts into the sky. Once you get thinking in terms of soil carbon, you'll never look at a harvested field the same way again. What might once have looked rustically beautiful and peaceful—quiet, golden fields; geometric ripples from the plow and combine—now seems, at least from a climate perspective, somewhat menacing.

Of the pair, Abe, in his late thirties, is the upbeat, can-do, jaunty one. He has pale blue eyes and smooth, fair, boyish skin, and likes to kid around.

At one point he kneels down in front of a freshly made hole and says, "That's what a worm looking upward sees—except that they're blind."

The two men mark out their grid with meter sticks capped with red flags. Peter lifts up a hunk of sod. As he digs deeper, the shovel makes a slurping, gurgling sound. "These soils are saturated," he says. I'm glad I borrowed those boots.

They do their work in earnest, Peter prepared with his knee pads, serrated knives, tin cans for the cylindrical shape, and an old-fashioned wooden classroom ruler. ("Top-grade state-of-the-art research here," he says drily.) He collects the soil samples in ziplock plastic bags and places them in a canvas sack with the words A NATION THAT DESTROYS ITS SOIL, DESTROYS ITSELF—the FDR quote that inevitably surfaces on T-shirts, publications, and email tags when you're dealing with soil champions.

I stand in the open field, feel the sun begin to withdraw its feeble late-fall warmth, and muse about what these guys are doing. Marking the land with posts and red flags, poking around in the dirt. And later, for Peter, the tedium of dealing with samples and data in his yellow school bus home. What's driving these two?

For one thing, they want to raise public awareness of soil's potential for absorbing carbon. According to Christine Jones, soils hold more carbon than the atmosphere and all the world's plant life combined—and can hold it longer, in a more stable form than, say, trees. She says that a soil carbon improvement of just 0.5 percent in the top twelve inches of 2 percent of Australia's agricultural land would effectively store that country's annual carbon dioxide emissions over the long term.

Here in the United States, Rattan Lal, of Ohio State, has estimated that globally soil carbon restoration can potentially store about one billion tons of atmospheric carbon a year. This means that the soil could offset about one-third of the human-generated emissions annually absorbed in the atmosphere. This is not some nifty trick, a way to cheat on nature; it's simply replacing soil carbon that has been lost over millennia. According to Lal, the carbon pools of most of the world's agricultural soils have been depleted between 50 and 70 percent. From man's earliest forays into agriculture about ten to thirteen millennia ago, that amounts to some fifty to one hundred billion tons of carbon.

"Restoring the depleted carbon pool in agricultural soils is essential to enhancing agronomic productivity for feeding the world population, and to improving the environment," says Lal. "It is a truly win–win option. The strategy of restoring carbon in world soils buys us time. It is a bridge to the future until alternatives to fossil fuel take effect."

Some folks make even bolder claims. Ian Mitchell-Innes, a South African rancher and trainer in Holistic Land Management, told me, "If we improve 50 percent of the world's agricultural land, we could sequester enough carbon in the soil to bring atmospheric CO_2 back to pre-industrial levels in five years."

Abe expresses it this way: "Worldwide, if the organic matter—which is about 58 percent carbon—in all the land that we currently farm and graze were increased 1.6 percent to a foot in depth, atmospheric CO_2 levels would be at pre-industrial levels. We'll have to do even better than that for many reasons, including if we want to get below three hundred parts per million of CO_2, since annual global carbon oxidation exceeds photosynthesis." He cites Allan Yeomans, author of *Priority One: Together We Can Beat Global Warming* and a longtime proponent of an agricultural solution to climate change, as inspiration for his soil carbon advocacy.

Whatever angle you take on this and whichever statistics you highlight, climate-wise the stakes are huge. Yet the notion of soil carbon has barely made a dent in our national or international conversations about climate change. And as Abe points out, even if we stopped emissions 100 percent (an occurrence that, given how global climate meetings play out, remains firmly in the realm of the hypothetical) but kept the same agricultural practices, we wouldn't be able to bring down carbon dioxide levels for a very long time. By itself, stopping emissions is insufficient. That carbon has to go someplace. As the Soil Carbon Coalition argues, it might as well go into the soil, where it can do some good.

We Become What We Measure

Peter Donovan has gray hair, blue eyes, and the serious demeanor of someone who broods over things like photosynthesis and the cycles of greenhouse gases and the folly of man. He's compact, not tall, but tall enough that living in 190 square feet is a squeeze. Still, he flatly turns down any offers of accommodations. "The bus is my home," he'll say, matter-of-factly. I sit in the bus and drink chamomile tea while he oven-dries (that is, he microwave-oven-dries) soil samples so as to measure soil density, a necessary factor in gauging the mass of carbon in the soil.

Peter's school bus has a small woodstove, a solar water heater, and a sawdust bucket toilet. There's a bed and sitting area, along with an Oriental carpet and a fine upright piano (Bach's "Englische Suite" is on the stand) and a reading nook (awaiting attention: the latest translation of *War and Peace*). "The thing I love about living here is the light," he says. "When you sit down, you have almost a 360-degree view." A little before six it grows darker and he puts on a headlamp so he can see his way as he moves around. That panoramic light fades, and the bus becomes a cozy room.

Peter is quick to note that he is not a soil scientist, which he sees as both an advantage and hindrance to his efforts: "It's hard to get funding." Because he's not associated with a known institution, he says, the coalition's work seems to make people uncomfortable. "I get asked, 'Are you with a university?' 'How do I know that what you're saying is true?'"

At the same time, he says, he has the freedom to ask questions and make connections that an institutional affiliation might not support. For example, the Soil Carbon Challenge measures carbon levels over ten years. Someone at a university, completing a PhD or seeking publication, has incentives to do research projects of no more than a few years. In government agencies and nonprofits, soil carbon work is geared to the "so-called carbon market." And all organizations—this is a pet peeve of his—tend toward fragmentation, so that soil conservation and climate mitigation are seen as separate, even competing, campaigns. All this means that stories that don't fit into a short time frame, aren't linked to profitable ventures, and/or can't be neatly tucked into departmental divisions may not get told.

It's a quixotic journey Peter's embarked on, seeing the country through the flat, broad windows of a bus, promoting ecological knowledge from outside the research and nonprofit establishment. And a somewhat roundabout way that he got here. After graduating from the University of Chicago, he worked in eastern Oregon herding sheep and cattle. In the 1990s, he trained in Holistic Management (HM) and met Allan Savory, the Zimbabwean rancher, wildlife biologist, and all-around maverick who developed the model. HM is based on Savory's observation that livestock can play a role in land restoration—that grazing animals, such as cattle, can be applied to land as a "tool" to improve it. (Savory officially enters our story in chapter 3, "The Making and Unmaking of Deserts.")

"Everything Savory said made complete sense to me," says Peter. Holistic Management became his intellectual base, informing his perspective on our advancing ecological plight. He traveled around the country and abroad, to Mexico and southern Africa, meeting HM practitioners, including Abe, and writing about his observations. Then in 2007, at a talk in Albuquerque, a researcher from the Natural Resources Conservation Service (NRCS), a government agency, told the audience that management made no difference to the accumulation of soil carbon and it was too hard to measure anyway. (That seems not to be the NRCS party line.) "That ticked me off," he recalls. "It was sort of a call to action. Abe and I had been kicking around a lot of ideas, so we started the Soil Carbon Coalition."

Abe, who has by now joined us after taking care of whatever mysterious tasks someone who manages a hundred-plus acres of land might have to do, grew up in central Vermont. In the 1990s, he lived in the community of Hard Rock on the Navajo Nation Reservation in Arizona where he studied and practiced agriculture, including approaches like permaculture and Holistic Management that have a regenerative component. "I did land restoration work with a group of Navajo men and women who started an organization called the Black Mesa Permaculture Project," he says. "There the basic line was: Our way of life depends on the land. It was 'common knowledge' that cattle and sheep, the way we were keeping them, were ruining the land, but when I talked to the elderly Navajo, they all said that the land was much better back when there were big herds and people herded the livestock in big, seasonal migrations." When he ran across Allan Savory's work using mega-herds of livestock to heal deteriorating land, places like Arizona's arid Black Mesa, which has been heavily mined, "the lightbulb came on in a big way."

Collins moved back to Vermont to apply these ideas to home turf. He keeps cattle and grazes them on land he manages—moving them from field to field according to Holistic Management principles, so that the fields are regularly but not overly grazed, trampled, and manured, then given time to regrow before being grazed again. For the most part their usual products, meat and milk, are incidental. Rather than, say, a dairy farmer or cattle rancher, Abe would be likely to describe himself as a grazier, a grass farmer, or even a carbon farmer.

They acknowledge that the day's soil measurements, as well as all those on Peter's cross-country swing, will do little to change land or minds. But this initial "baseline tour," as the Soil Carbon Coalition calls it, begins the process of putting together data: the Holy Grail of science-based environmental advocacy.

"Getting carbon baseline measures is not a practical thing to do," Peter concedes. "But we need real data. Not because people will make decisions based on that—because people don't necessarily do that—but if we're going to make a shift toward considering soil carbon a key to land function, we'll need a place to stand, for example in tons of carbon gained or lost per hectare per year over a decade. Right now there's not a whole lot of data, especially in terms of changes over time. The thing about data is that it can change our beliefs about what is possible and not possible. It can work on our imaginations, which is important because in our society we have a lot of conflict between the defenders of the impossible, those who say, 'We can't do it,' and the artists of the possible."

"We're talking about building topsoil," Abe breaks in. He's restless. He wants to fix something, now. At the very least, he wants me to understand the vast implications of soil carbon: that bolstering soil carbon means adding organic matter to the soil, which means building topsoil—the precious growable layer of the ground we ply, where the biological magic of food and forage occurs. It's been accepted as a given that topsoil forms over geological time; the figures generally bandied about are that it takes five hundred to a thousand years to generate an inch of new soil. But this refers to the slow weathering of rock—nature left to its own devices, without human interference. Actively working to stimulate soil formation—through maintaining ground cover, increasing biological activity, imposing levels of disturbance that add oxygen and moisture, as can be done with livestock—can accelerate the process. That is, really speed it up, to an inch or more a year. In regenerative agricultural circles, it's been done. (We'll see how in chapter 2.)

"Accelerated topsoil formation is the great work of our time," Abe says, as in a proclamation. "It's the centerpiece for addressing the environmental security and economic development issues facing all of us in one fell swoop. Now, is there an opportunity to build topsoil, or are we stuck with a thousand years needed to build an inch? In any complex system, you need to measure. So that's where we start."

Enter: The Carbon Cycle

In a conversation with Peter, one phrase that often comes up is *managing toward.* What he means is that in any management situation, there's a difference between "managing for" something and "managing against." He believes that one reason we—that is, we humans, or at least we modern, Western humans—continually back ourselves into a corner is that our impulse is to manage against. As one example, our approach to medicine is managing *against* disease rather than managing *for* health. This has helped bring us to a situation where we need increasingly bigger guns to fight off infection while a huge chunk of the population suffers from chronic disease or unnamed malaise. Managing for health, by contrast, would emphasize diet, exercise, avoiding toxins, and building immunity (chapters 5 and 7 discuss how soil plays into these factors).

He believes that this is why our carbon problem—the need to do something about rising stores of carbon in the air—leaves us stuck. We're trying to manage against tossing more carbon into the atmosphere. What we can and should be doing, Peter argues, is to manage for a carbon cycle that does more work: splits more water and carbon dioxide and results in more water-holding, fertility-enhancing soil organic matter. Our current approach, which worthy and well-meaning environmental organizations have signed on to, has been to plead with would-be polluters to stop polluting. As a friend of his wryly put it, this strategy amounts to: "Let's wreck the world more slowly." This is a valid effort compared with the alternative—protecting nature where we can is essential—but hardly a route to lasting ecological soundness. However, Peter believes that by a shift in our thinking, we will start asking different questions and emerge with different solutions to the problem of carbon levels run amok. What we need is to tell the story of our predicament in a different way and keep ourselves focused on the notion of *managing toward.*

In his various stops along his travels by bus—he's done about fifty-four hundred miles when I catch him in St. Albans—Peter gives occasional soil carbon workshops. Since Bennington is on the way to the Boston area, his destination after Vermont, I arrange for Peter to give a workshop here in town, which the One World Conservation Center, a relatively new environmental education center and reserve, housed in an old Hojo's, kindly offers to host. I am expecting a rehash of what I've already learned, all the good news about soil carbon and why we should promote the building of soil carbon. I could sit back and occasionally nod knowingly. But no, this is a full-on theoretical lecture on the carbon cycle, and the historical forces that have interfered with our ability to see this universal, ongoing process—a mainstay of any high school science class, if handled in a perfunctory way—as a key to environmental restoration.

I can hardly hope to do justice to this tour de force on the history of scientific thought, but I'll give it a go. Actually, I'll do it twice: First, I'll highlight some ideas that surface in our sweep through the natural sciences. Then I'll do the bumper sticker version.

Peter starts by posing a question: Where does most of the mass of plants come from? Plants grow in soil, so most people think the answer is soil. *Wrong!* This was proved in the seventeenth century by Jan Baptist van Helmont, who grew a willow tree for over five years and found that the weight of the soil it grew out of had hardly changed. Water? That's what the Flemish van Helmont thought. Nearly a century and a half later, a Swiss scientist, Nicolas-Théodore de Saussure, refined this experiment by enclosing plants in glass and monitoring the water and carbon dioxide given to them. He demonstrated that carbon in plants—the basis for plant organic compounds—was obtained from carbon dioxide in the air; the hydrogen in these compounds came from water. This set the stage for an understanding of photosynthesis, a plant's ability to take in solar energy and combine it with water and carbon dioxide to make food and mass.

Photosynthesis is one of those things we all kind of know about, but Peter wants us to appreciate just what a powerful force this building-plants-from-carbon-dioxide trick is. "Life doesn't just sit there," he tells the group. "It *does work*. Nature is a process, not a collection of things." We tend not to think of biological processes as work because they're always going on in the background, slowly and quietly, at least to the extent that we can observe. If we define work mathematically, as force over distance, day in and day out the work of photosynthesis exceeds the total of the world's industry by a factor of nine. Plants, then, move many times more carbon molecules than does the burning of fossil fuels.

Now we're ready to take a look at the carbon cycle:

The carbon cycle is what we generally call the cycle of life: birth, growth, death, and decay, in which photosynthesis plays an important part. Photosynthesis uses sunlight to split water and carbon dioxide, and builds compounds that become food, fuel, and biomass. The reverse reaction is respiration, oxidation, or combustion—which in the 1780s French scientist Antoine Lavoisier recognized as similar processes. This other side of the cycle turns carbon compounds back into carbon dioxide and water and releases energy.

When we look at the carbon cycle in the context of soil, this is what we see: If carbon-rich soil organic matter is being oxidized by common bacteria in the presence of oxygen, the soil is losing carbon. This creates a negative feedback loop, with soils losing more carbon and water and plants struggling to grow. If, however, we've got more carbon in the soil, we've got fertile ground for plants, more photosynthesis, and a *positive* feedback loop that takes carbon dioxide out of the air. In our biological system, soil functions as a hub, or intersection, for the carbon cycle.

Starting to make sense?

At this juncture we switch gears and consider our carbon dioxide problem—that "inconvenient truth" of having too much carbon dioxide in the atmosphere, due in no small part to human activity. Understanding of this problem has to a great extent derived from the Keeling Curve, which was featured in Al Gore's landmark documentary on climate change, depicting the concentration of carbon dioxide in the atmosphere. It is named for Charles David Keeling, a scientist at the Scripps Institution of Oceanography at UC San Diego, who began gathering carbon dioxide data in 1958. Since then, readings have regularly been taken at Mauna Loa in Hawaii. It is arguably among the world's most recognizable scientific images.

Keeling's emblematic diagram depicts carbon dioxide as a continuously rising line, with slight fluctuations to account for seasonal variations. For example, the line dips when it's summer in the Northern Hemisphere, since carbon dioxide is drawn from the atmosphere by newly growing plants; during the North's winter, with the seasonal decay of crops dying back and leaves decomposing, the rate of emissions ticks up again. The visual image of that slender line, relentlessly climbing upward on into the temporal frontier, can only evoke in the observer a feeling of anxiety and helplessness if not outright dread. If I drew up a graph of what happens to my blood pressure while looking at the Keeling Curve, it would look very much the same.

As Peter sees it, the message we take from the Keeling Curve is this: Carbon in the atmosphere traps heat. This is a problem of atmospheric pollution. The cause of this is the burning of fossil fuels. Therefore the solution is to reduce fossil fuel emissions. The implication—and the conclusion we accept—is that by curtailing emissions we will be rewarded with a moderating curve. However, according to the Intergovernmental Panel on Climate Change (IPCC) Fifth Assessment Report, if, in 2007, we completely stopped emitting carbon dioxide, it would take nearly a century to bring concentration levels down to 350 parts per million.

In other words, we learn that the way to deal with our carbon dioxide problem is through technology (low-carbon energy sources; whiz-bang contraptions to capture carbon), political will, or both: strategies that aren't getting us anywhere, create divisions among people and groups that should be working together, and, says the IPCC, won't have much near-term leverage on atmospheric carbon dioxide.

What if, and this is a big "what if," rather than wringing our hands over technology and politics, we looked to biology? And rather than flinging ourselves at the inexorable incline of the Keeling Curve, we tapped into the ongoing work of the biosphere, the carbon cycle? By burning fossil fuels, we're basically short-circuiting the carbon cycle by using concentrated stores of energy created by photosynthesis long ago. What if we could work with natural processes, increasing photosynthesis and slowing down the oxidation of soil organic matter?

Peter likes a quote from Dwight David Eisenhower: "If a problem cannot be solved, enlarge it." He relates this to our carbon problem: "We have to get beyond the question of pure climate and ask what makes the earth tick. And the answer is, the carbon cycle."

Now that you've humored me through the last several paragraphs, here's the bumper sticker equivalent: OXIDIZE LESS, PHOTOSYNTHESIZE MORE. Feel free to stick it on your (yes, I know, carbon-dioxide spewing) car.

More Work from Nature, Less from Tractors and Plows

Bringing our story back to the visit to St. Albans, I return to the bus the next morning, carrying a round of pastries from the Cosmic Cafe. I pass up the monster, skeleton, and pumpkin cupcakes, each under a colorful solid inch of frosting. (This is one town that takes Halloween seriously.) Before the last turn in the road I see a group of brown cows, shades spanning from tan to near black. These are among the cattle that Abe moves from pasture to pasture grazing according to a dynamic plan, four-legged tools for boosting carbon in area soils.

Peter is organizing his traveling bag. He's off to Saskatchewan to teach workshops and set up some more baseline plots. Abe's driving him to the airport in Montreal, about an hour and a half away. After Canada, Peter's plan is to travel south ahead of the bad weather, with scheduled stops in Massachusetts, southern Vermont, Pennsylvania,

Maryland, Virginia, and North Carolina and on through Texas and the Southwest before making his way back to Enterprise. "The bus is not a place for cold weather," he says.

Abe gets milk from the mini fridge, wary of the ceiling as he maneuvers. Peter smiles. "Of course I checked the height before I bought this bus," he says. "My son is six feet and it doesn't work for him."

I ask Peter to reflect on his journey so far. "I'm 210 days in, and nothing too bad has happened," he says. "I've had a number of visits from the police, just checking me out. Twice in Iowa, and once I was just about to cross the Mississippi into Illinois. I was camped at an information center. A cop came by and checked my ID, which interrupted my phone conversation.

"It's been my intention to go slow, get to know some people and see some country. I'm a pretty experienced traveler—I've done thousands of miles on horses and mules. I saw a bunch of innovative farmer-grazers working toward soil health in a dedicated fashion. That was encouraging. But it was dismaying to drive through three or four states and see nothing but corn and soybeans." Large industrial farms break the heart of people like Peter who care about soil carbon. Generally after harvest the land stands uncovered and the carbon in the soil oxidizes.

"Sometimes I think what I'm doing is crazy," he admits when I ask. Yet he continues on for the opportunity to talk to people about the intimate connection between the soil and the sky, and the potential for healing this represents. "The opportunity we have to enhance the carbon cycle is a human opportunity, not a technological opportunity," he says. "It's easy to get sidelined by technology, but it's about people: our beliefs, assumptions, and what we think is possible. Most of what we hear about carbon and the global carbon cycle is threatening and negative. It's a bad situation and we don't seem to have much power or leverage over it. All our environmental and economic concerns depend on the ways carbon and water move—and water follows carbon both into the air and into the soil. We need to understand that human decisions have an enormous influence on the way these cycles function."

In order to get this point across, he has a new prop, which I later see on his visit to Bennington: a protest-style placard that reads: OCCUPY THE CARBON CYCLE. This is, after all, the autumn of 2011, the heyday of Occupy.

I see that Abe Collins is looking at his watch. But he isn't urging me to leave just yet. "Every major advance in human economic and social life has been tapping a new carbon source," he says. "Think of farming, oxidizing organic matter to release nutrients to grow crops, think of forestry—for timber and charcoal, think of coal and then oil. We've tapped the really rich energy sources. Now we're at a point where we've tapped the soil. It's tapped out. We've lost an estimated 50 to 80 percent carbon in our soils over the last 150 years. We've got to get people to make the fundamental link between our economic and ecological cycles." And the importance of recharging the soil, to keep the ground covered and get a broader range of plants growing again. At least on that score, he believes he knows what needs to be done: "Let the animals and nature do more of the work."

Cows, Methane, and All That Hot Air

If I mention that I'm writing a book that deals with climate change and anyone catches the word *cows* in the title, the first reaction is usually, "Oh, you mean it's about how *cows are causing global warming!*" For some reason this myth has caught the public imagination. Let's get this clear: The burps and farts of bovine animals are not to blame for climate change.

First, to understand how the rumor got started:

1. Cattle, like all ruminants, emit methane as part of their unique digestive process (from the front end, actually). According to the EPA, ruminant livestock annually generate about eighty million metric tons of methane, which is approximately 28 percent of the global methane emissions attributed to human-derived activity.
2. Methane (CH_4) is a greenhouse gas. In terms of heat trapping, it's about twenty-five times more potent than carbon dioxide.
3. Therefore, we should cork those cows and then we won't have such a problem with greenhouse gases.

This makes some degree of sense, right? At least until we look at cows-and-methane within a biological context:

1. That methane-is-twenty-five-times-more-potent figure is widely reported and accepted by many as fact. However, David Mason-Jones, an Australian journalist and author of *Should Meat Be on the Menu?*, says this metric "simply does not compare apples with apples." CO_2 and CH_4 have different weights, with a carbon dioxide molecule nearly three times as heavy as a methane molecule. Rather than comparing the global warming potential of a molecule of carbon dioxide with a molecule of methane, the twenty-five number expresses the activity of a kilogram of methane versus a kilogram of carbon dioxide.[2]

   Plus, methane in the atmosphere breaks down much more quickly than carbon dioxide; in the presence of oxygen CH_4 turns into CO_2 and H_2O, or water.
2. As Mason-Jones points out, the methane cycle functions as a subsystem within the larger carbon cycle of growth, digestion, and decay: "The production of methane in a landscape should be recognized as an inevitable consequence of plant growth because, when plants grow, they inevitably get eaten by something . . . Methane is one of the biological consequences of this and will occur whether there are farm animals in the landscape or not."
3. It's not the cattle themselves that are making the methane. The gas is a by-product of enteric fermentation, the type of digestion ruminants perform, and released by bacterial symbionts in a cow's stomach chambers in the process of breaking down cellulose. Humans can't eat grass, but cattle, thanks to their microscopic partners, are performing the service of turning inedible grasses into protein that we can eat.
4. There seems to be little correlation between methane levels and the number of ruminants. A joint 2008 report from the FAO (the UN's Food and Agriculture Organization) and IAEA (International Atomic Energy Agency) noted that since 1999 atmospheric methane concentrations have been stable while the population of ruminants worldwide grew at a rapid rate, raising the question of whether livestock play much of a role in the greenhouse gas situation.
5. The amount of methane cattle generate has a lot to do with how livestock is managed. According to Andrea Malmberg, director of research at the Savory Institute, when cattle are in crowded feedlots (concentrated animal feeding operations, or CAFOs) the manure is handled in liquid form, in lagoons or holding tanks—an ideal scenario for the production of methane. If, however, cattle graze on fields in lowered concentration, the manure becomes fertilizer. She adds: "These are human decisions, not the cows'."

Biochar and Its Source (Fire)

If you scroll around on the topic of soil carbon, sooner or later you're bound to bump into biochar. Biochar is charcoal used in an ecological context, primarily as a soil amendment, created by burning plant material (wood and crop residues) under low-oxygen conditions, known as pyrolysis. (It's also sometimes known as terra preta, or "black soil," after research in the Amazon showed that indigenous people there applied it to their soil more than a thousand years ago. It is stable, concentrated carbon that, when added to soil, enhances fertility and sequesters carbon. Advocates tout biochar as the answer to climate change. (*The Guardian* referred to proponents as "charleaders.") For me, such grand claims triggered the too-good-to-be-true alert. So I contacted Steven Apfelbaum, a world-recognized expert on ecological restoration and the founder and chairman of Applied Ecological Services in Wisconsin. He's my go-to guy whenever I need a scientific reality check. Here's what he said:

> Biochar is one of the ways nature has stored carbon. In most forest soils, there are chunks of it: coarse fragments of carbonized wood. Lots of it is easily 400 to 1000 years old. You'll find this in northern forests, Midwest oak savanna systems, Coastal peatlands, the Pacific Northwest, all the way to up the Arctic—carbon stores, stable and protected.

Historically, the primary origin of biochar is wildfires, he said. According to Joel S. Levine, a senior research scientist at NASA, about 30 percent of global annual carbon dioxide emissions can be attributed to biomass burning.

Apfelbaum continues:

> Today most wildfires—perhaps 90 percent—are human-ignited, and many occur under extreme conditions. Uncontrolled, catastrophic wildfires lead to major erosion, opportunistic species on scarred land, and greenhouse gas emissions—not to mention any other losses involved. However, there are also prescribed fires that can be administered as part of restorative management. One byproduct of a controlled fire is biochar. Another is greater land productivity. Carbon is sequestered in the soil and you have a more open, less competitive, environment conducive to plant growth.

> Many forests have way more trees than the land can support. The soil is subject to erosion, and any biochar there is not stable. As the ecosystem deteriorates the soil carbon and biochar oxidize, combining with atmospheric oxygen to form CO_2. Controlled burns can restore soil to stable condition and stimulate regrowth of grass, sedges, and other plants. The kind of burning I'm talking about [is] light ground fires to trim overgrown tree canopies that could otherwise be prone to wildfire. With prescribed burning, you slowly morph the ecosystem back to a healthy condition—rather than administer the near-death experience of an all-out wildfire.
>
> I see prescribed burning as a low-hanging fruit for the biochar industry. That and using waste wood and other cellulose material to create carbonized products for agriculture. We should be using nature as a role model and look at conditions that have led to this and learn from that rather than fabricating products.

Right now, says Apfelbaum, biochar production is primarily experimental and small-scale, though he believes it holds promise.

One sometimes cited concern about biochar as an industry is that agricultural lands, particularly in the third world, might be turned into charcoal plantations, diverting millions of acres from food production and creating a new breed of monoculture. Plus, can we assume that commercial biochar will have the same soil-enhancing effects as that formed in nature? At least one study casts doubt on this. Then there's a question of whether the production of biochar could generate more emissions than assumed, including the black soot that causes respiratory disease. Sounds like we should tread slowly with biochar. After all, it is playing with fire.

Chapter Two

Carbon Trading: Nature's Version

> Soil may be considered as the conversion of rock by two processes. One is a process of aging, the other is a process of living.
>
> —P. A. Yeomans[3]

> Carbon trading is something that has been going on for millennia in our soils. It underpins the health of our whole ecosystem.
>
> —Christine Jones

ANY EARNEST DISCUSSION OF SOIL CARBON will inevitably wend its way to Australian soil scientist Christine Jones. Over the last eight to ten years, Jones's articles and posts on her website—www.amazingcarbon.com—have become foundational documents for those, like Peter and Abe, who seek to rein in carbon dioxide levels by working in tandem with ecological cycles. Her determination to promote soil restoration as an engine for environmental renewal is matched only by her frustration that the powers-that-be seem set on ignoring her message.

In Australia, as in the United States, the conventional belief is that what determines the crop-worthiness of a patch of land is its "nutrient status," the levels of the macronutrients nitrogen, potassium, and phosphorus. These can be bumped up via synthetic and organic fertilizers. For Jones, it's all about soil carbon, which needs to be built by biological processes and is hindered by the very amendments purported to guarantee bumper yields. She's taken this message on the road, talking to farmers, gardeners, and politicians willing to listen, attending field days and giving workshops across Australia and abroad, telling it straight. Her presentation aids tend toward the earthy: yellow buckets, a tape measure, tufts of grass weighed down with soil and roots.

A forthright woman in her late fifties who looks most comfortable out in the field under a tan, broad-rimmed Akubra hat, Jones has had her share of dustups with the agricultural and academic establishments. There was, for example, "the great salinity debate" of the early 2000s, over how to address high salt levels that were undermining farmland productivity in much of the country. To national farming and environmental groups, the answer was to plant lots of native trees. According to Jones, though, prior to European settlement the bulk of the continent was grassy woodland rather than actual forest, and dryland salinity signaled problems with the soil, specifically a lack of soil organic matter. She's proposed and launched several carbon sequestration programs, only to hit the inevitable bureaucratic wall. Yet she has helped bring *soil carbon* into national parlance in Australia—something that's yet to happen anywhere else.

When asked about her early background, the phrase Jones offers is *paradise lost.* In part, she means this on a personal level. She describes an idyllic childhood among loving family in a sleepy fishing village on New South Wales's picturesque South Coast, living in a tiny log cabin graced by broad, leafy tree ferns. "We were largely self-sufficient," she recalls. "I have fond memories of going fishing with Dad, helping Mum make jam, tending 'my garden' (Dad had given me a section of my own), walking to the dairy across the road for our daily billy of fresh warm milk, dipping my fingers into the thick layer of delicious cream on the homeward trip. Much to my mother's dismay I had a great fascination for frogs, lizards, spiders, and snakes, which I kept in special places in the garden." When she was ten years old the family uprooted and moved to Sydney. Life in the city was not her style: "It was an alien world beyond my comprehension."

Jones refers also to the loss of Australia's paradisiacal wildness, much of which she's witnessed in her lifetime. The lush, fertile countryside, filled with fabulous creatures (fabulous at least to the nonphobic and unsqueamish), has been tamed, engineered, and exploited, as much of Australia's land has been overgrazed, paved over, or had the life farmed out of it. When Europeans first settled in the early 1800s, the soil was described as peaty, crumbly, and soft—so soft that horses were prone to stumble and occasionally broke their legs. Trouble was, the loamy veneer was thin and the climate hot and dry. Yet the colonials went at the land with their plows and tills as if they were still in moist, temperate

Europe. The introduction of foxes and rabbits, which killed and outcompeted native species like dunnarts (a small marsupial related to the Tasmanian devil) and the spotted quoll (another marsupial), didn't help.

Today Australian land is largely arid, acid, and saline. In October 2011, *Rolling Stone* ran a cover story on "the end of Australia," a nightmarish piece that portrayed the country's struggle with heat, droughts, fires and floods, dying reefs, and parched riverbeds as a harbinger of the ravages that climate change has in store for the rest of the world. While many in Oz would dispute one countryman's characterization of their homeland as "the petri dish of climate change," most would acknowledge the island nation's ecological fragility. Upward of 70 percent of Australia's land is considered degraded. The country has the highest carbon emissions per capita of any in the world.

Jones went on scholarship to the University of New South Wales, graduating with First-Class Honours with a major in wool and pastoral sciences. While a degree in wool science might sound like the punch line of some Aussie joke (just add a reference to beer, convicts, kangaroos, and a "G'day, mate!" or two), keep in mind that wool was long a core national industry; it's said that Australia was "built on the sheep's back."

Jones's subsequent work as a textile technologist is what prompted her interest in soil. She explains that there are "several processes involved in turning raw wool into yarn. First, the wool is scoured, or washed, then carded and combed into tops, which are like coils of rope. The wool tops are then spun into yarn. If wool fibers are uneven in thickness along their length, they break at the narrowest point. This can result in a great deal of wastage during the top-making process and can also reduce the 'spinnability' of the yarn. Variations in the thickness of wool fibers are mainly due to changes in the quality and quantity of pasture throughout the year. To cut a long story short, my interest in wool became an interest in pastures, which became an interest in soil, which ultimately became an interest in the question, 'what makes soil tick?'"

Jones went back to the university for a PhD, exploring the biochemistry of the plant–soil relationship. She says: "I quickly realized that plants are anything but 'passive' and soils are anything but 'inert.' There is this extraordinary world beneath our feet, fueled by energy from the sun, of which we are largely unaware. For me, this realization was like the lifting of a heavy fog. It suddenly became clear that within the soil lay answers to myriad so-called problems."

Not only in the wool industry, but other agricultural sectors. And over time Jones came to believe plant–soil dynamics provided answers to larger environmental concerns, among them our surfeit of greenhouse gases and the daunting challenge of growing food for a seven-billion-and-counting world. The question that's continued to haunt her, from her stints in the business world and the Commonwealth Scientific and Industrial Research Organisation (CSIRO), Australia's main research agency, to her current role as independent scientist, has been: How can she draw the needed attention to soil health when there are entire industries and institutional hierarchies set up to peddle Band-Aid solutions while perpetuating the underlying problems?

Fueling the Soil Battery

Christine Jones is the type of thinker who speaks to the gaps between chapters in a textbook, concepts not covered on the test. Fortunately, she has a vivid way of explaining these ideas, with a touch of the clarifying metaphor. For instance, she looks at the mechanisms of photosynthesis from the standpoint of soil carbon, making the connection between what happens on the leaf and what happens in the ground. Here we move from Peter Donovan's near-cosmic atmospheric scenarios to the nitty-gritty of the mundane workings of a clump of soil. We all dutifully read our high school science textbooks so we know that through some miraculous alchemy, the chlorophyll that speckles a plant's leaves is able to transform air and sunlight into energy the plant can use. Then what happens? The plant grows. The plant energy becomes food for us and for other creatures, large and small. In time, the plant dies and the remnants of that plant energy biodegrade and return to the soil as organic matter. What Jones says is that there's *another* process going on, one with important implications for sequestering carbon and building topsoil, and doing so rapidly—thus hitting two angles of our carbon problem. She calls it "the liquid carbon pathway."

To understand this evocative phrase, we need to slightly revise our conceptualization of a plant. Rather than a mostly green thing embedded in the ground that pulls water and nutrients from the soil, Jones would have us think of a two-way pump. The upward flow is water, minerals, and other substances the plant needs; the downward flow is soluble carbon (dissolved organic carbon) that seeps into and out through the plant's roots so as to feed other organisms in the soil. This downward carbon flow stimulates the production of humus, the organic component of soil that is a repository for carbon as well as the basis for fertility. The more carbon in the soil, the more energetic the microbes; the more energetic the microbes, the more mineral particles are broken down; the more minerals are broken down and made available to plants, the more humus. "We have the idea that plants *take* from the soil," says Jones. "Nothing can be farther from the truth."

We know that through the process of decay, plant material humifies and becomes a stable, carbon-rich component of soil; it's not prone to chemical or physical breakdown and, if undisturbed, can remain in this form for centuries, even millennia. Jones is saying that the *living plant* also creates humus, serving as a conduit for the energy that drives its production. In doing so the plant is delivering the sun's energy underground, where it is stored as living soil. In Jones's words, the soil "behaves like a battery, fueling the above-ground ecosystem from the soil up as well as from the sun down." The soil, then, serves as a storehouse for latent energy—in the form of humus—pulled down from the sun.

As Jones recognized when she was a young scientist, soil is far from inert. At any given moment there's much ferrying of food and energy going on beneath our feet—an intricate system of reciprocal and contingent exchanges, the complexity of which would put those geniuses who invented stock derivatives to shame. For our purposes, let's keep things simple and focus on how the liquid carbon pathway makes its way to one particular life-form essential to the fixing of carbon in the soil (and, not coincidentally, Christine Jones's favorite biota), mycorrhizal fungi.

These are fungi that forge symbiotic associations with the roots of plants. Under natural conditions, upward of 80 percent of plants that flower have cooperative links with mycorrhizal fungi. This includes grasses—though, curiously, not brassica, the vegetable family of broccoli, kale, cabbage, and the like. Mycorrhizal fungi have long, threadlike filaments called hyphae that carve paths into the soil, extending the "reach" of a plant and increasing its access to nutrients and water. These hyphae can attain impressive lengths: A teaspoon of soil in the vicinity of a robust plant might hold strands that, when stretched end to end, would span nearly half a mile. According to one study, the hyphae found in the top four inches of forty-three square feet of healthy grassland would run the length of the equator.

A key function of mycorrhizal fungi (the name means "root fungus") is governing the nutrient give-and-take between plants and soil, acting as a trading center for subterranean biochemical barter. The mycorrhizal fungi "provide essential nutrients to plants in exchange for liquid carbon," says Jones. "They cannot utilize any other form of energy other than dissolved organic carbon that has been provided by a living plant. In the absence of plant roots, you cannot grow mycorrhizal fungi in a laboratory." They also enhance a plant's resistance to disease, drought, and difficult soil conditions like acidity or salinity. Jones likens the long, stringy hyphae, full of sugar they've taken from the plant, to "fairy floss" (that's Australian for "cotton candy").

The liquid carbon the plant discharges into the soil via its mycorrhizal fungal partners also feeds other microbes that, in turn, supply minerals to plants in a soluble form they can use. Plants happy, microbes happy, soil lavished with minerals and organic matter. A plant with mycorrhizal connections can transfer up to fifteen times more carbon to soil than a non-mycorrhizal counterpart.

There's another aspect to these root fungi scientists are still learning about: glomalin, a sticky secretion—it's been called "soil's superglue"—that coats the spindly hyphae. This glycoprotein (both a carbohydrate and a protein) was only discovered in 1996 by Sara F. Wright, a soil scientist with the USDA's Agricultural Research Service (ARS). Glomalin is significant for two reasons: It holds carbon, storing it for as long as several decades (a study by microbiologist Kristine A. Nichols of the ARS determined that glomalin represents on average 15 percent of carbon in soils); and it binds soil particles to create aggregates, which lends soil its tilth—that soft, granular quality you get when you run a handful of good soil through your fingers. This helps keep soil stable and resistant to erosion while allowing for air and water flow. As one USDA brochure asks rhetorically, "Does glomalin hold our farm together?"

Kristine Nichols, who works out of the ARS's Northern Great Plains Research Laboratory in Mandan, North Dakota, says this is not hyperbole: Glomalin may be central to soil formation. The "stickiness" of glomalin—like "little gobs of chewing gum," she says—binds together particles of sand, silt, and clay and plant and other organic debris to form aggregates within the "string bag" of the stringlike hyphae. The glomalin then forms a "latticelike waxy coating" that keeps the aggregate intact and protected and ensures that water and nutrients are available to the associated plant.

"Soil structure is the 3D structure of open and closed spaces," Nichols explained in an email. "If the closed spaces are formed by soil aggregates rather than individual soil particles, the open space between these aggregates will be larger, which will allow for better water and gas movement along with reducing compaction."

In other research, Nichols found more glomalin in prairies marked by native plants as compared with non-native plants, reflecting the interdependence that has evolved among indigenous grasses and mycorrhizal fungi.

Mycorrhizal affiliations benefit a plant in terms of growth, resistance to drought, and immunity to disease (they release antibiotics to fend off invading organisms). The fungal networks also enhance the soil structure, which means better air and water circulation and root support. There are now lines of mycorrhizal products, such as Fungi Perfecti's MycoGrow, available to farmers, gardeners, and landscapers.

All good—but it's important to know that mycorrhizal and synthetic brews don't mix. "Mycorrhizal fungi are inhibited by the use of water-soluble nitrogen and/or phosphate fertilizers, herbicides, insecticides and, of course, fungicides," says Jones, noting that for this reason, "we do not find them in abundance in conventional agriculture." In crop fields left fallow, without any plant cover, the mycorrhizal network literally starves to death. They're also disrupted by tilling; stick a shovel in the soil and you're likely to fray those delicate tendrils. In a kind of biochemical treadmill, a grower may be adding fertilizers and other chemical agents to bolster a crop that's languishing mainly due to the loss of mycorrhizal associations. Reflecting the fungi's role in drawing carbon into the soil, Christine Jones says that when fungicides are used, "carbon flow will be stopped before it even starts."

How much carbon can be brought into the soil and stored? According to Jones, "under appropriate conditions, 30 to 40 percent of carbon fixed in green leaves can be transferred to soil and rapidly humified, resulting in rates of soil carbon sequestration in the order of five to 20 tonnes of CO_2 per hectare per year." If we wish to "revitalise all terrestrial life forms, including people," she says, the way to do so is to restore the soil battery. This means creating the conditions for the liquid carbon pathway to flow uninterrupted in the soil, giving biology the chance to do its thing.

Soil Making—The Speedy Version

In talk of global environmental crises, we hear a lot about carbon dioxide, but we don't hear much about soil. However, international organizations, including the UN, are beginning to address global soil degradation as a threat to food and water security. Globally, each year some seventy-five billion tons of soil is lost. That would cover about thirty-eight thousand square miles of arable land, an expanse larger than the nation of Austria. The phrase *peak soil* has begun to appear in media reports. In the United States, Cornell soil scientist David Pimentel says that 90 percent of our cropland is losing soil to wind and water erosion at thirteen times the rate that soil is being formed. Given that we depend on these vanishing soils to support a growing population, the pace of erosion and land degradation has become a matter of increasing concern.

To Christine Jones, sequestering carbon and creating soil are the same thing; by maximizing the liquid carbon pathway, you're kick-starting the process of building soil. This single insight frees us from the assumption that soil generation is something that can only occur over geological time. The belief has been that topsoil isn't something one can just "make." Therefore, the best we can do is to try to preserve what we've got and somehow bide our time for the five hundred years needed to grow a measly inch of the stuff. Conventional wisdom is that soil organic matter forms in the top layer of soil, the product of degrading biomass. True, this uppermost stratum of soil is highly biologically active, with decomposer fungi breaking down the organic residue; this mechanism both stores carbon and releases carbon dioxide into the air. However, thanks to the liquid carbon pathway, the mycorrhizal fungi, in partnership with their botanical hosts and via the carbon-holding properties of glomalin, are also making humus. This means of generating soil organic matter emits far less carbon dioxide—the transfer of carbon takes place *underground.*

This alternative route to soil building is significant for a few reasons. First, the carbon that accumulates near the soil surface, say in the first four inches, says Jones, is mostly "short-chain, labile carbon, indicative of rapid turnover." This "restless" carbon is critical to the soil food web, the community of organisms that reside in the earth's upper stratum. With the carbon in flux, however, it's not so effectively stored. The more deeply the carbon can be stowed away, the more securely it is preserved and the greater its contribution to soil structure and function over the long term. This highly stable carbon is enhanced in the lower soil profile via the liquid carbon pathway; the surface carbon, most of which oxidizes, rarely makes its way to the stable lower substrate.

Also, it calls attention to biological dynamics occurring at depth, which, says Jones, are generally ignored. The Kyoto Protocol, for instance, only addresses carbon sequestration down to thirty centimeters (one foot). In an article called "Carbon That Counts," passed around soil circles like a 1970s rock bootleg among music fans, she writes: "Routine assessments of agricultural soils rarely extend beyond the top 10 to 15 centimeters and are generally limited to determining the status of a small number of elements, notably phosphorus (P) and nitrogen (N). Overemphasis on these nutrients has masked the myriad of microbial interactions that would normally take place in soil; interactions that are necessary for carbon sequestration, precursor to the formation of fertile topsoil."

At depths below a foot soil does form—more rapidly than in the soil's upper horizons, Jones contends. Here's how: The liquid carbon compounds that are pumped down by the plant bolster the stores of organic matter. At the same time, the liquefied carbon stokes the earth-dwelling microbes, which get to work dissolving the mineral portion of the soil. These minerals at once feed the plant (remember, the mycorrhizal fungi are busy making mineral exchanges down there at root level) and build humus at a fast clip. In Jones's words: "A large proportion of the liquid carbon is humified. That is, it's transformed to humus, a stable form of soil carbon that has amazing properties, not the least of which is the rapid conversion of compacted subsoils into soft friable topsoils, dramatically improving soil depth and function."

You got that? One farmer I met, Gene Goven of Turtle Lake, North Dakota, put it more succinctly: "You build soil where the roots go—*down*!"

The humus, formed at lower depths, advances the goal of carbon sequestration because it is more stable. It's what Jones characterizes as "carbon that counts" in that it's more resilient than carbon in the top layer, the level that's typically measured. It doesn't disappear in a drought, the way surface organic matter, formed by decaying biomass, is more prone to do. It retains minerals and water so that these are accessible to plants as needed, and therefore creates the environment for better food production.

With the plant's carbon pump driving the process, humus-rich topsoil forms *downward*, well beneath the surface. That is, as long as the mycorrhizal networks stay intact. "The positive feedback loops render the liquid carbon pathway somewhat akin to perpetual motion," says Jones. "You can almost see new topsoil forming before your eyes." Since the liquid carbon pathway streams dissolved carbon through a plant's root system, all this is happening beneath the usual threshold of observation. So according to standard carbon modeling, this process doesn't exist.

If It's Not in the Textbook, Can It Be True?

Sequestering carbon dioxide and generating high-quality topsoil in one go? This sounds great. There are just two problems, both of which cause Jones—a woman who wants to get things done—no end of aggravation: One is that the scientific/agricultural establishment doesn't believe it. The other is the touchy matter that the chemical additives beloved by many government institutions and industry have a tendency to mess it all up.

Let's tackle the second one first.

If the first transformative revelation for Jones was that soil was a crucible for so many agricultural and ecological challenges, the second was the realization that those entities with the power to do something about soil—large agricultural services companies, whether government- or privately funded—didn't really want to know about it. In a sense this was another "paradise lost," in that she worked in conjunction with educational and governmental agencies for more than two decades under the assumption that they shared the goal of improving the land and its ecological viability.

> The great learning for me was to understand the implications of the enormous industry that depends on us *not* finding solutions to problems in agriculture. These so-called "problems" generate income for the ever-expanding ancillary industries including the manufacture of synthetic fertilizers, herbicides, insecticides and fungicides. Far greater profit is derived from developing, manufacturing, marketing, transporting and applying toxins—to the food we feed our kids—than is made by farmers. Indeed, the major portion of farm income is now expended on the inputs required to maintain production as soil function fails.

The emphasis on nutrients like nitrogen and phosphorus, she says, has obscured the importance of mycorrhizal and microbial interactions that create soil and sequester carbon. Not only that: The presence of high levels of nitrogen from fertilizer sends a signal to plants to *reduce* the supply of liquid carbon to microbial symbionts, effectively inhibiting the microbial associations that would otherwise supply atmospheric nitrogen for free.

As time went on Jones became outspoken about topics like CSIRO, Australia's preeminent scientific research body, forging alliances with (in other words, accepting money from) such corporations as Bayer CropScience, which sells genetically modified (GM) canola. This won her few friends among agricultural power brokers. She told Australian Broadcasting Corporation's news show *Landline*, "People, sometimes nicely, say I'm a lone visionary. When they are not being nice, they say other things that I won't repeat." Her views on soil building and what she refers to as "the microbial bridge" are dismissed due to "lack of data," and her research is dismissed because the results confound conventional beliefs.

Industrial expediency is one factor; lack of confirming documentation is another. One reason people aren't paying much attention to mycorrhizal and microbial activity, says Jones, is that government- and industry-funded research sites tend to be located on conventionally managed land, which doesn't exhibit natural plant–soil dynamics. "In most of today's agricultural soils, the sequestration pathway is not there. The microbial bridge—the biological processes that build soil carbon and humus—has been blown by chemical additives."

Now the other problem: No one believes it's possible to build topsoil this fast.

As an example, let's leave Christine Jones and head a few hundred miles southwest toward Gulgong, a former gold-mining town in the Central Tablelands of New South Wales, near which Colin Seis runs a two-thousand-acre farm. The farm, Winona, has become Australia's Exhibit A for accelerated soil creation. Fifth-generation farmer Seis is now as well known for "pasture cropping," the land-management model he helped develop, as for the (gnarly, stout) merino sheep and (winsome, long-eared) kelpie working dogs his family breeds. This farming technique has turned out to be an ideal approach for rapidly building soil.

Pasture cropping, a method of sowing annual cereal crops into perennial (preferably native) pasture, was initially devised by fellow fifth-generation farmer Darryl Cluff, who describes the technique in his 2003 book *Farming Without Farming*. When Cluff assumed management of the family farm, Olive Lodge in Birriwa, near Gulgong, the land was in a sorry state. His quest to figure out how to restore the farm led him to examine Australia's farming history—and seek ways to reclaim what was lost.

"Our early European history records that Australia was a country with vast areas of grasslands, as well as huge areas with scattered trees," says Cluff. "It seemed to me that the grassland community became so efficient at survival that they were out-competing the trees." These native grasses, however, were depleted by decades of intensive grazing; they were seen as having little value and were often replaced by non-native annuals.

"The trouble is that these [native] plants are not well understood, the seed is very difficult to harvest and the plants have specific requirements for their establishment," says Cluff, who helped form an organization called Stipa Native Grasses Association, which focuses on practical applications of native grasses. "It takes a very long time for quality native pastures to naturally re-establish after a cropping phase. However, our native pasture plants were always there, the last feed available during drought and the first to respond to the rain when it finally came. I thought that they had tremendous value, but they didn't quite fit in with our farming system. My thoughts were on how to grow profitable crops without the cost of re-establishing pastures."

So he left the pasture there. In 1995, after an eighteen-month drought that only added urgency to these questions, Cluff direct-drilled an oat crop into a pasture of dormant native redgrass (*Bothriochloa macra*)—and despite the scant moisture, the crop thrived. A year later, Seis began pasture cropping, too. When Cluff chose to concentrate on his own farming business—due in large part to criticism from the agricultural bureaucracy for his departure from standard practice—Seis began promoting pasture cropping among other farmers.

And Winona became the showcase. The two-thousand-acre property has been in Seis's family since the 1860s, when his great-great-grandfather started a sheep and wheat operation. Seis's father, Harry, took over in the late 1920s and grew wheat conventionally (with lots of heavy machinery) and prospered. That is, until the soil began washing away. To address the degraded soil conditions he turned to government-subsidized superphosphate fertilizer and was extremely successful, particularly during the postwar wool boom. By the 1970s and '80s, the soil was acidic, the water table increasingly saline, and weeds were taking over. In 1979, after a bushfire that destroyed most farm buildings and killed three thousand sheep, Colin dispensed with the fertilizer that had gone up in price and was now costing him tens of thousands of dollars each year. He turned his attention to the native grasses—redgrass, spear grass, kangaroo grass, wallaby grass, and others—that flourished once they weren't competing with phosphate-loving plants.

When you look at a pasture-cropped field at maturity you see the crop, row after row of grain standing tall in tufts. But rather than emerging from a brown base of tilled soil, the plants are rising from a bed of green pasture. The original concept was to find a cheaper way of sowing oats to feed the animals. But what Cluff and Seis stumbled upon was a means of re-creating the biological community of native grasslands. In natural grasslands annual and perennial plants coexist, each benefiting the other. Broadacre farming, which removes existing ground cover so that crops are sown on cleared fields, damages soil structure, interrupts fungal and microbial associations, and releases stored carbon. Pasture cropping, by contrast, leaves soil dynamics intact. It also supports a variety of plants, in particular many deep-rooted grasses engaged in carbon–mineral–water exchanges underground.

At Winona, Seis saw a difference after two years. The land developed better resistance to drought, a greater diversity of plant, insect, and microbial species, higher soil carbon levels (a 200 percent increase over ten years), and an overall improvement in ecological function. Productivity improved, since the perennials reseed (so there's no need to sow). And the perennial ground cover means constant protection from wind and water erosion.

Pasture cropping "mimics the function of natural grassland, where perennial and annual species grow symbiotically," says Seis. "It may be that the direct drilling of the cereal crop into pasture when it is dormant provides food for the bacteria and fungi at a time of year when natural pasture is not active. At the same time, managing pasture for native perennial grasses means they can grow to full expression. This way they maintain soil structure and store sufficient root reserves to be involved in nutrient trading." The only downsides to pasture cropping are the need to buy or alter machinery and that crop yields might dip, particularly at the outset. But increased profits from combining pasture and crop production on the same land should more than offset this.

As for soil building, over a ten-year period the depth of topsoil more than quadrupled—from four inches to eighteen inches. Christine

Jones, who has compared soil at Winona with soil on adjacent paddocks, notes that this was during a period of below-average rainfall in the region.

Seis expresses concern that among science organizations in Australia, "most innovative land-management practices are not taken seriously. Recently the NSW Department of Agriculture measured soil carbon increases on my farm 'Winona' which showed an increase of over 160% to a depth of half a meter. They are now disputing their own data and are not going to use this data in their overall research in carbon sequestration using pasture cropping techniques, even though the site on my farm is the second-oldest 'pasture cropped' site in the world." Seis says pasture cropping is now being practiced on two thousand farms and ranches around the globe.

In one instance, Christine Jones saw her own research used to "prove" that landowners could not "afford" to focus on building soil carbon because, according to five top-ranking CSIRO scientists, the necessary nitrogen, phosphorus, and sulfur supplementation would be too costly. The rationale for the CSIRO paper was that such inputs would be necessary. (One of the authors used this assumption to question Colin Seis's significant increases in soil carbon at Winona.) What these scientists failed to take into consideration was that in high-carbon, high-functioning soil with the liquid carbon pathway intact, soil microbes are able to fix nitrogen and release phosphorus and other elements, making them available in a plant-accessible form.

"Institutional soil scientists, funded by agrichemical companies, are doing their utmost to prevent this information being accepted because the humification process (and hence the storing of the sun's energy in the soil) does not proceed where there are high levels of chemical inputs," says Jones. "Once farmers 'get' this, the big end of town (in the ag world, at least) will have nothing to sell. Farmers will not want to use toxic chemicals because their use results in soil degradation—which is a symptom of the loss of soil energy."

On a more hopeful note, she adds: "Paradise has been lost on much of our agricultural land, but I know it can be regained."

The idea that soil can accumulate rapidly is not new. Charles E. Kellogg, a professor of soil science with the Natural Resources Conservation

Service, wrote in 1949, "Some people speculate about how much time is required 'to build an inch of soil material.' The answer could well be, 'somewhere between 10 minutes and 10 million years.'"

One tool that reportedly allows for faster soil building is the Keyline plow and design system, originally developed in the 1940s by P. A. Yeomans, a farmer and engineer (and his son, Allan Yeomans, who wrote *Priority One*, which inspired Abe Collins). The chisel-shaped plow decompacts and aerates the subsoil with minimal disturbance; water can infiltrate and conditions improve for fungi and microorganisms. With Keyline plowing and planned high-density grazing, Yeomans was reportedly able to produce four inches of humus-rich soil in three years, starting with bare, sandy ground.

"You can build soil fairly quickly," says Courtney White, co-founder and director of the Quivira Coalition, a nonprofit based in Santa Fe, New Mexico, devoted to the economic and ecological vitality of western working landscapes. He has visited several Australian farms, including Winona. "With caveats—it depends on the rain, worms, and condition of the land. Most folks think that with soil, it's all downhill. Ranchers are building soil. But they don't think about it in the climate context. They want the grass to grow. The challenge is how to scale it up. For building soil, what's important is not so much the weathering of rock and minerals but the biological processes. When we think in terms of chemistry or geology it's one thing. If you think of biology, it's very different."

This echoes Yeomans's understanding of building soil: "It merely has to convert the sub-soil into fertile soil. The length of time that this takes is related to the life cycles of the life in the soil." To quicken these life cycles, the Keyline system combines the deep-reaching chisel plow with grazing timed so as to stimulate plant growth.

On a visit to North Dakota, Gene Goven showed me some before-and-after slides that portray changes on his land since he implemented Holistic Management (his story is in chapter 6). "Here we had hardpan because of tillage," he said, meaning that the top layer of soil was compacted and impervious to water. "The roots only went down three to five inches. After 90 days of soil building we had ten inches of aggregate structure. We built six inches of topsoil in one season, soil that didn't exist in the spring. How many years does it take to build topsoil? Five hundred? We build soil down—that's the paradigm shift. It can be done. I'm doing it."

Abe Collins combined using the Keyline plow with Holistic Planned Grazing in northern Vermont, and says that in one year they went from eight inches of topsoil on top of gray clay to sixteen inches of topsoil. "We used the Keyline plow in the sequence that Yeomans discovered changes subsoil into topsoil: graze the grass, subsoil to a few inches below the current topsoil layer, allow for regrowth, graze again. The 'soil conversion phase' of Keyline is usually suggested to last for about three years. In this time, one would subsoil at least once per year. The soil changed as predicted by Yeomans and other practitioners around the world who have worked with Keyline Soil Formation. We subsoiled ground twice that year, and that seemed to push the soil system to a new state."

When I was up in St. Albans, I saw this soil as Abe dipped his shovel and grabbed a brown-black fistful and let it roll over his palm. But I didn't see it six years ago, when he started his soil-bolstering regimen. Which is precisely the stumbling block: Anecdotal reports and fence-line comparisons are fine for practitioners sharing land-improvement strategies, but getting beliefs about soil creation to budge takes a body of evidence. As Peter Donovan says, to date "there's not very much good data, no longitudinal study to measure changes over time." This, he notes, was one motivation for pursuing the Soil Carbon Challenge, since increasing carbon and building soil are facets of the same process. Without the numbers, it's too easy for agency bureaucrats and industry reps to say "we need solid data" and to dismiss these ideas as "pie in the sky" and individual claims as "tall tales." It takes a long time to gather the figures. Unfortunately, we don't have that time; the problems these methods could potentially solve are pelting us now.

Managing for Photosynthesis

With all these approaches to building soil and building soil carbon—the Keyline system, pasture cropping, Holistic Planned Grazing (the focus of the next chapter), or, most commonly, a combination of these—what's key is the pace of photosynthesis. Barring interruptions to natural processes, the more photosynthesis is occurring per acre of land, the more carbon is being stored and the more soil is being made. The exchanges facilitated by mycorrhizal fungi, the fulcrum of what Christine Jones calls "the microbial bridge," prime the plants for higher rates of photosynthesis. Another factor that promotes photosynthesis is greater leaf-surface area. Which means that any time plants are removed, as in field clearing or burning, photosynthetic capability is diminished.

Remember our bumper sticker—OXIDIZE LESS, PHOTOSYNTHESIZE MORE? That's what's happening here—the plants are taking in more carbon dioxide and storing more carbon.

As for the "oxidize less" part, the thing to avoid is bare soil: Without plant cover, soil carbon is prone to bind with oxygen and go airborne. The plant overlay also buffers temperature and slows evaporation, establishing conditions for microorganisms to thrive, thus galvanizing plant growth, which in turn stimulates photosynthesis. Soils depleted of carbon are oxidation risks. With inert soil the liquid carbon pathway runs into a dead end; rather than entering the soil food web, the carbon oxidizes and returns to the atmosphere as carbon dioxide.

Which does our current mode of agriculture favor? By planting annuals and plowing and clearing so that the soil is open to the elements for much of the year, we're managing for oxidation instead of photosynthesis, regardless of whether we realize it or not.

"We create so much bare soil in conventional agriculture that a lot of the sun's energy is dissipated as heat rather than converted to biomass," says Jones. "Every kilogram of glucose produced via the photosynthesis process represents 16 megajoules of sunlight energy bound in a biochemical form. If that same amount of light falls onto bare ground rather than onto a green leaf, the energy is radiated back to the atmosphere." Peter Donovan likes to refer to spots of bare ground as "sunshine spills," creating a parallel with an oil spill that might occur when a pipe leaks or something otherwise goes wrong. The potential of the energy has been lost.

Depending on what you put on and in the soil, you're managing either for photosynthesis or for oxidation. These tracks are self-reinforcing; you're launched onto one trajectory or another: toward carbon sequestration and soil health or toward carbon depletion and soil degradation. For example, more leaves mean more photosynthesis. Robust leaves also mean more roots, which means more carbon streaming into the soil. Which means, at once, greater fertility and more carbon sequestered. Steering land on a positive course can happen very quickly; Jones says changing "from annual to perennial groundcover can double levels of soil carbon in a relatively short period of time." In addition to uninterrupted photosynthesis, perennial plants are continually pumping carbon and playing host to mycorrhizal and microbial activity. (Provided those perennials are managed appropriately, which is the subject of the next chapter.)

Bare ground sets in motion a different set of circumstances: moisture loss, less food for microbes, and therefore less microbial activity, which results in fewer nutrients for crops, leading to less-than-vigorous leaves, less photosynthetic activity, and less carbon flowing through the roots and being cached in the soil. Each contingency perpetuates the downward spiral. Says Jones: "When carbon sequestration stops, soils lose structure—like decomposing wood. And, like a dead tree, soil no longer functions as a living, growing thing."

But we needn't head down that road; at every juncture there's a choice, and for every mechanism there are two ways to look at it. As Jones says, "Every tonne of carbon lost from soil adds 3.67 tonnes of carbon dioxide to the atmosphere. Conversely, every one tonne increase in soil organic carbon represents 3.67 tonnes of CO_2 sequestered from the atmosphere and removed from the greenhouse gas equation."

As for the arithmetic of food, Jones says, "According to those who are good at math (which doesn't include me!) the planet's current vegetative cover captures around 3000 EJ of the sun's energy. This photosynthetic capacity would appear to be sufficient to feed only 3.3 billion people. The other 3.7 billion people, then, are being supported by photosynthetic energy captured in past eras, embodied in fossil fuel." By restoring the capacity of the world's soils, however, we can do better. To offer one example, Jones notes that Colin Seis's land can carry twice as many livestock animals per acre as neighboring land. "In other words, it can feed twice as many people."

Nitrogen and Soil: A Cautionary Tale

Nitrogen (N in the periodic table) is essential to all life, and forms the basis of proteins, genetic material, and enzymes. Availability of nitrogen is often a limiting factor in growing crops. Not for lack of nitrogen per se: Nitogen gas, N_2, comprises nearly 80 percent of the atmosphere. But in its gaseous form, N_2 is a highly stable molecule that plants can only access in certain ways: via the breaking of the N_2 bond through lightning and the element's subsequent descent by rain; or through the activities of nitrogen-fixing bacteria in the soil or on the root nodules of legumes. Plant-usable nitrogen is also present in manure and humus, but is released slowly so is, well, limiting.

All of this changed in the early twentieth century with the discovery of the Haber-Bosch process, which converts nitrogen gas to nitrogen-based fertilizer on an industrial scale. The manufacture of nitrogen fertilizer was a key component of the "Green Revolution" beginning in the 1960s. (With a slight detour into nitrogen-derived explosives, like nitroglycerin and TNT.) Crop yields soared, enabling the dramatic growth in world population, and, in this country, cheap food. Thanks to modern chemical innovation, we seemed to have broken the Malthusian barrier: the notion that population growth will always exceed expansions in agricultural output.

Because no caution was taken, this becomes a cautionary tale. By the 1990s, the increase in yields hit a wall. Particularly in the United States, where low fuel costs meant fertilizer was cheap, the response was to add more. The Law of the Minimum—that a plant's capacity to grow is determined by the scarcest nutrient, popularized by Justus von Liebig, the nineteenth-century scientist who pinpointed nitrogen's role in allowing plants to thrive—became the Law of the Maximum: better to put on more fertilizer than risk a bad crop. Globally, the amount of nitrogen applied to agricultural land is rising about 15 percent a year.

Global dependence on nitrogen fertilizers has led to several problems:

1. Soils reach a nitrogen saturation point, and the excess either oxidizes to become nitrous oxide (N_2O), a potent greenhouse gas, or leaches into water as nitrate. In any body of water this can lead to algal blooms and hypoxic zones. In sources for drinking water, nitrate poses a health hazard.
2. It had been thought that fertilizers bolstered soil carbon because rapid plant growth meant more decaying biomass. However, research is finding the opposite, that synthetic nitrogen depletes soil carbon. It speeds up growth of

microorganisms that feed on nitrogen at the expense of other soil dwellers, and these turbocharged microbes proceed to consume the humus.

3. Long-term use of nitrogen fertilizers acidifies soil, which leaves plants vulnerable to diseases and pests, particularly those that flourish in acid environments. Acidic soil also interferes with a plant's ability to take up minerals and nutrients, including, ironically, nitrogen.
4. Synthetic nitrogen fertilizer creates in the soil a classic addiction scenario: Because the soil is depleted, it can't function without help. The only thing that does the trick is another hit of nitrogen. This masks the underlying problems, even as the soil is further compromised.
5. The manufacture of chemical nitrogen fertility relies heavily on fossil fuels (including, in places like China, dirty coal). On a conventional farm, production of the fertilizer it uses accounts for nearly a third of its total energy consumption.

The use of artificial nitrogen fertilizer short-circuits the biological nitrogen cycle. In so doing, says Christine Jones, "It wreaks havoc on the organisms that make soil fertile. Adding synthetic nitrogen inhibits the microbial associations that would otherwise enable plants to access some of the 78,000 tons of nitrogen that sit above every hectare of land. However, if farmers and gardeners support microbial activity in the soil, there will be no need for inorganic chemical fertilizers."

Can those who grow our food forgo the lure of fast nitrogen? It will be a challenge, since today the production of these fertilizers involves an interconnected complex of the agricultural, energy, chemical, and explosive industries, bonds proving every bit as tough to crack as that of the original N_2 molecule.

Chapter Three

The Making and Unmaking of Deserts—The Grazing Paradox

We are a desert-making species.

—Elisabet Sahtouris, evolution biologist

OVER THE PAST THREE DECADES, French filmmaker Yann Arthus-Bertrand has observed the world's landscapes from above. He's captured aerial scenes from more than a hundred countries, racking up thousands of hours of helicopter time. These lofty glimpses of nature's edges and crevices left him in awe of not only the earth's beauty but also its fragility; this made him an environmental activist as well as an artist. Even when depicting devastation or abject poverty, the images are stunning: In one shot a woman in Burkina Faso picks cotton as nearby a child, naked and with belly distended, rests amid the white fiber as though nestled in a cloud. Another depicts the Alberta tar sands. The whorls of viscous bitumen make a beautiful abstract design, like wood grain.

One aspect of our "altered planet" Arthus-Bertrand has become quite vocal about is desertification, the march of land degradation in dry areas. He has one arresting photo of a caravan of camels making its way across twilit ocher sands, its shadow a ghost train of itself riding the crest of the dune. This was taken near Nouakchott, the capital of Mauritania, a nation bordering the Sahara that's been particularly affected by desertification. The caption notes that prior to becoming the capital in 1960, the area was a fertile grassy plain; now a city of more than two million, it has "the desert at its door."

I met the filmmaker briefly at the United Nations in September 2011, at the first United Nations General Assembly devoted to addressing the topic of desertification. He spoke at the press conference that followed, as a goodwill ambassador for the United Nations Environment

Programme (UNEP), along with Hifikepunye Pohamba, the president of Namibia, and Luc Gnacadja, of Benin, the executive secretary of the United Nations Convention to Combat Desertification (UNCCD). The presser was sparsely attended, unnervingly so. In the way we have of sizing up our tribe, I glanced around and counted maybe half a dozen reporters. The discussion was as much about the apparent lack of interest in the topic as the topic itself. Gnacadja, a former architect who, particularly for a UN dignitary, gives an impression of youthful affability, began his remarks by saying, "If this were about climate change, the room would be full."

The UNCCD is kind of a poor cousin among the three conventions established at the Rio Summit in 1992, the other two being the Convention on Biological Diversity (CBD) and the United Nations Framework Convention on Climate Change (UNFCCC), which receive far more ink and attention. But desertification is hardly a fringe problem. Here, courtesy of the UNCCD, are some sobering facts:

Drylands—the arid, semi-arid, and sub-humid areas with seasonal, often unpredictable rains—are complex, delicate ecosystems that though resilient are vulnerable when land and water are not sustainably managed. Drylands account for 41.3 percent of the world's landmass, including 44 percent of land under cultivation. Each year upward of twelve million hectares (thirty million acres) of productive land are lost to desertification; this means an area the size of South Africa is slipping away each decade. With two billion people living in drylands, mostly in the developing world, the loss of arable land is pushing large populations trying to eke out a marginal living into poverty, starvation, and migration. Today 1.5 billion people depend for their food and livelihoods on land that is losing its capacity to sustain vegetation. It's estimated that half of armed conflicts can be at least in part attributed to environmental strains associated with dryland degradation. Desertification has contributed to the downfall of civilizations: think Carthage, Mesopotamia, ancient Greece and Rome.

Desertification is not a "natural" development—the result, say, of dune movement or shifting sands. It is a man-made process, driven by actions that disturb the life cycles of many plant and animal species that have adapted to dryland conditions. These endeavors include overcultivation, deforestation, poor irrigation design, poor livestock management, and the use of technology ill suited to the landscape. Land degradation is not unique to drylands; actually, most land deterioration occurs in areas you wouldn't consider dry. The problem is that in drylands the margins that determine adequate rainfall or water scarcity, a good harvest or crop failure, are exceedingly thin. Which means that if you're in England or the U.S. North Atlantic coast, you can get away with messing things up somewhat without facing immediate consequences; residents of Australia or North Africa lack that luxury. Weather-related events like droughts and floods do not in themselves cause but can intensify and hasten the decline to desert—land that becomes "dirt" with life neither on nor within it.

It's ironic that desertification might, global-crisis-wise, need to compete for airtime with climate change and biodiversity loss—because they're all connected. Here's a hypothetical snapshot of how it plays out: Degraded soils in arid lands lack the capacity to store carbon; whatever carbon is in the soil oxidizes to form carbon dioxide, contributing to climate change. Absent plant cover and moisture, bare ground absorbs heat (if you've scorched your feet on hot sand, you'll know what I mean). If you remove plant cover and litter over one square yard, the soil heats up and the microclimate is altered. Do that over the whole of North Africa or most of Australia, and you've changed the macroclimate. The added heat makes any moisture present more prone to evaporate, and want of moisture leaves the soil less than welcoming to microorganisms, the absence of which deprives plants of nutrients and other beneficial exchanges. With the habitat unfriendly for growth, the range of plants that can survive is limited, which then limits the birds and insects that might pollinate or spread seeds. Sparse vegetation means little protection from winds or heavy rains, so these occurrences lead to more soil blowing or streaming away, more carbon oxidizing.

So intertwined are these three—desertification, climate change, and biodiversity loss—that we can consider them manifestations of the same problem: The biological cycles underlying life on earth have been thrown out of whack. We can't hope to make inroads on any one of them without addressing all of them. However, this is not how it's usually discussed.

The word *desertification* and its association with droughts and barren landscapes makes us think that we're dealing with a "just add water" situation: that all those benighted regions need is a good dose of rain and the land would burst into tulips and daffodils. But if we were to reduce the problem to its essence, desertification is really about soil: soil losing its "aliveness," the wherewithal to sustain life.

One reason desertification does not rank higher in our awareness is that it's often thought of as a third-world problem. It happens *over there*, in regions we're unlikely to visit, places many of us haven't heard of let alone be able to pronounce. It's the kind of thing we'd read about on what used to be called "the Africa page" in the newspapers, shake our heads in momentary pity, and think *Thank God I don't live there*, assured that something like that could never happen where we live. In fact, the continent with the highest proportion of its dryland areas classified as severe or moderately desertified is North America, at 74 percent. In the European Union, another region you might think immune, thirteen countries suffer some degree of desertification.

While working in the United States in the early 1990s Allan Savory, the African wildlife biologist and rancher who developed the planned grazing framework called Holistic Management (its application is generally called "Holistic Planned Grazing"), made this observation: Most experts on desertification attribute its development to overpopulation, overstocking of livestock, cutting down too many trees, poverty, war, lack of education and/or technology, and the overexploitation of shared resources. In West Texas, where Savory was working, none of these factors was present. The rural population was declining, as were livestock numbers, while mesquite trees grew undisturbed. There was no war, no dearth of money, technology, or education. Nor could one blame the "tragedy of the commons," since all the land was privately owned. Yet West Texas was desertifying as rapidly as anything he'd seen in Africa: Sand dunes were forming, rivers were drying up, and the water table was dwindling.

Clearly something else was going on.

When it comes to desertification, few have done as much innovative thinking about it, or been as roundly criticized for it, as Savory. A key idea underlying Holistic Management—that grazing animals can serve as a tool for preventing or reversing the desertification process—has offended many scientists and academicians, even as thousands of practitioners attest to the improvements it has brought to their land. Here was someone outside the agricultural and scholarly mainstream who did his research in the open air and presented his improbable-sounding conclusions in plain language. Many footnote-laden articles and spirited diatribes (including more than a few that have landed in my inbox) have been devoted to the cause of proving Savory wrong. So far the best these folks have been able to do is agree to disagree with him.

Knowing the polarizing figure he is, I was surprised when I met Allan Savory in New York—where his ideas would be the topic of a Deepak Homebase conversation—to encounter a slight, older man with alert eyes and a gentle manner and who had a way of pausing an instant before committing a thought to speech. I even felt a tinge of familiarity upon hearing him talk. It was the accent: My husband, Tony, is from South Africa, and it still touches his speech. As a child, Tony was a keen observer of animals and wanted to be a game ranger when he grew up. Of course he knew of Savory's work long before I did.

The evening with the Deepak Chopra folks was a bit odd. The livecast panel discussion was held in a showroom at ABC Carpet & Home; the chairs and sofas we sat on—all contemporary and very stylish—bore price tags. The "curated conversation," I felt, veered too much into the metaphysical to be informative. I could swear I saw Savory grimace when talk turned to a shift in higher consciousness.

But the appearance at such a venue was yet another sign of Savory's changing fortunes in terms of how his ideas are received. In 2003 he was given Australia's Banksia International Award "for the person or organization doing the most for the environment on a global scale," previous winners of which include Rachel Carson and David Attenborough. In 2010 his Zimbabwe nonprofit, Africa Centre for Holistic Management (ACHM), won the Buckminster Fuller Challenge Prize, a prestigious award granting $100,000 to a project with "significant potential to solve humanity's most pressing problems." Also that year ACHM received a $4.8 million grant from the United States Agency for International Development (USAID) to expand its work in Africa. In recent years he's presented at numerous international events, including UNCCD's Land Day 2011 in Bonn, Germany. In a phone interview, UNCCD's Luc Gnacadja called Holistic Management "a game changer" for addressing desertification. There's even a bit of glam, as actor Ian Somerhalder, star of *Lost* and *The Vampire Diaries*, has announced plans to make a documentary about Savory. Somerhalder told *E! Entertainment News*, "The goal of the movie is to win Allan Savory a Nobel Prize for agriculture, which has never been done."

Let's zero in on what Savory discerned about the process of desertification, and how the means to address it may be contrary to what we might think. First, how his ideas evolved.

Savory grew up in Bulawayo, Rhodesia, in a family with long roots in colonial southern Africa. As Sam Bingham wrote in *The Last Ranch: A Colorado Community and the Coming Desert*, Savory was "the distilled essence of British colonial history," raised "amid more pith helmets and khaki than one could find in a shelf of Kipling." In the mid-1950s, after completing a degree in biology and botany at the University of Natal in South Africa, he worked for the Colonial Service in what was then Northern Rhodesia (now Zambia). As a wildlife ranger—at twenty, he was the youngest ever taken on for this position—his task was to stop poachers and kill rogue elephants and man-eating lions. He mentioned this to me so casually that I had to pause and repeat, "Man-eating lions?" Apparently, with people living in proximity to lions and the habitat of the big cats and their prey under stress, lions may hunt humans. Once a lion has killed and eaten a person it must be killed, since it will have developed a taste for our kind. The most dangerous man-eater Savory encountered had done in thirty-five people.

Lion trapping in the wild led to an important skill: tracking. "I went out with native trackers, but all too often they would lose the track on getting close," Savory says. "Finally I realized: they were afraid of finding the lion. So I knew that in order to do this work I was going to have to learn to track myself. If you're going to follow a lion, it does no good to just shoot any lion. You need the one experienced with killing people, because it will do it again." He learned to read the land, to peruse his surroundings for tiny clues—a disturbed leaf, a piece of grass at a slightly unnatural tilt, a broken spiderweb—as to what animals might be fleeing or on the prowl, interpreting the signs to try to understand what was going on. The next time a native tracker said he lost the track, he'd be able to do it himself. "Liebenberg wrote of tracking as the origin of science," he says, referring to the South African author of *The Art of Tracking*, adding, "It could well be true."

Too Many Animals . . . Or Not Enough?

As commander of the Tracker Combat Unit during Zimbabwe's long war to gain independence, Savory also spent thousands of daylight hours looking at the ground and subsequent nights reflecting on what he saw. "At night we couldn't light a fire because then we could be found," he said. During the dark nights at camp sleeping on the ground—"never the luxury of a tent"—he pondered and tried to interpret all he'd observed during the day. Why was it easier to track people over land devoted to certain purposes and harder over others? What were the variations in soil exposure, litter, and plant life, and what caused such differences? He also began to understand that the greatest danger to wildlife was not poaching but habitat destruction—the same thing that ultimately threatened humans. He reasoned that if a few animals were killed by poachers either to feed their families or to sell, the animal population bounced back. But it couldn't bounce back once habitat was destroyed.

Savory was stationed in areas to be set aside as future national parks, in the Zambezi and Luangwa Valleys. He recalls: "I was looking at the habitat all the time and saw that it kept getting worse. Some land was healthier. This was mostly in pristine wild areas where herds of buffalo, elephant and other game along with large numbers of lions and other predators still existed. But even there it was deteriorating. I began to speculate that we game rangers and biologists were greater dangers to the animals than the poachers. This was not a popular statement."

One way the wildlife service sought to protect the land in order to create a preserve was by "resting" it: removing and resettling the local population. "We lied like good bureaucrats," says Savory. "We said this was because there was sleeping sickness in the area, though the population had been living with that for centuries. As the land got protected, it deteriorated. We were starting to lose species like bush-buck and nyala and large areas of reeds and rambling fig bush were disappearing. There seemed no conceivable explanation for this except that there were too many animals. Like most scientists I interpreted the data I gathered on land degradation to fit my beliefs. I concluded that there were too many elephants."

Privately, Savory was still asking the question: Why, paradoxically, did wildlife habitat lose vitality—showing signs like erosion, bare soil between plants, patchy growing patterns with some spots overgrazed and some overgrown with fibrous, woody vegetation, all indicative of incipient desertification—even as the game rangers took strong measures to protect it? In his official capacity as a research officer in the game department, he felt compelled to act and so, in a move that pains him to this day, he wrote a report recommending the culling of elephants.

"The idea of culling large numbers of animals in preserves was shocking to the government," he says. "They formed a committee of respected ecologists to verify my reports and inspect the worst areas. I brought the people into the field with me to see the condition of the land and so we went ahead and shot thousands of elephants."

The result? Says Savory: "The problems got much worse."

Desperate for answers, while consulting at a ranch he chanced to pick up a South African farming magazine from the coffee table. He read a piece by a botanist with an unusual but apparently successful sheep-grazing method that revived land—and was causing a row. He was intrigued enough to drive five hundred miles to Middelburg in the Eastern Cape to meet John Acocks.

The older man attributed the area's degrading land to domestic animals' tendency to selectively graze, eating the grasses they preferred until only the shrubby, less tasty plants were left. This was in contrast with wild roaming herds, as each of several grazing species would favor different plants. To counter selective grazing, Acocks devised a system that concentrated the animals so they'd eat everything in one spot, after which he'd move them to another. This was contrary to the widespread belief that unless sparsely placed in minimal numbers, domestic animals were bad for the land. In one corner of a paddock under Acocks's supervision, Savory dropped to his knees and dug his finger in the soil. As he later wrote in *Holistic Management: A New Framework for Decision Making*, "Here I saw domestic livestock could do to the soil what I had seen with large herds of buffalo. The surface was broken; litter lay everywhere; water was soaking in rather than running off; aeration had improved; and new seedlings grew in abundance." Savory could tell this patch had been trampled, which clarified a few points for him: Healthy soil is contingent on the actions of the animals that live on the land; livestock could be made to act upon the soil as wild herds had done; and animal disturbance could benefit as well as harm a landscape.

The next piece of the puzzle arrived in the form of a book called *Grass Productivity* by André Voisin, a French scientist. Voisin made the observation that overgrazing is determined not by numbers of animals but by length of exposure. In other words, one cow grazing a field at will can do more damage than, say, an entire herd that nibbles on the grass for a brief time and then moves away. In the former case, edible grasses are bitten to the ground; in the latter, the grasses have a chance to recover. Savory realized that managing for time was important, as was managing to ensure for the high physical impact of hooves. In this way he began to integrate the work of Acocks and Voisin together with his own observations.

The centrality of timing helped Savory understand why the land set aside for animals deteriorated: The human population living in the bush had been keeping the animals "wild." The local people used to beat drums to protect their gardens along the river, and had muzzle-loading guns to hunt and safeguard their crops. He explains: "When these lands became protected as future parks the game became more sedentary. Once we removed the people, animals came along to the rich soils by the water and they could linger all day in the shade of the trees. In the parks we had big sturdy trees for shade, plenty of nutrition and water. There was no one banging drums or shooting rifles, just biologists and binoculars and tourists with cameras. The animals stayed. They were there too long. They needed to be moving."

When Savory realized that livestock were the best tools to reverse the very land degradation they were blamed for causing, he faced a dilemma. For thousands of years pastoral people had herded livestock with great knowledge of their animals and land, yet had caused the great man-made deserts. And for the last hundred or so years range scientists had developed a host of different rotational grazing systems, only to see desertification increase—as Savory had himself observed on ranches and at research stations. How, then, were livestock to be managed? It had to be in a manner that could account for the daily complexity of livestock needs, wildlife, crops, other uses of the land, and variable weather. Because no one had ever pursued such an approach in the biological or range science fields, he looked to other professions for ideas. Of these, military planning in battlefield conditions offered the most hope.

He turned to Britain's Sandhurst military college, whose planning process had been used by the Rhodesian army. "I took the simple *aide memoire*, a military planning idea, and adapted it to planned grazing of livestock integrated with wildlife, cropping and other land uses, and to accommodate constantly changing weather and other conditions, such as a fire sweeping across the land," he says. "This rapid planning process had been refined over hundreds of years in which men are trained to simply concentrate on one small part of the situation at a time, and to develop the best possible plan right now in simple steps, each building on those before. All that had to be added was to record each step on a grazing chart because the military had not needed to address time, volume of forage, sizes of land, behavior of animals and more. This planning process results in the many different factors having a bearing appearing on the chart over the months planned, then allowing the moves of the livestock to be plotted so that the animals are in the right place, at the right time, for the right reasons and with the right behavior. The military is geared to stressful conditions just as farmers are always stressed."

In the 1960s and '70s over many trials and, as he admits, much error, Savory worked to develop a model for land restoration using livestock. During this time he also ran a consulting business and a farm and game ranch, formed an elite tracking combat unit during Rhodesia's long civil war, and served in the Rhodesian parliament, where he led the opposition to Prime Minister Ian Smith's racist regime. In 1979, after being denounced by Ian Smith as a traitor and threatened with detention, he went into exile; years before, a Molotov tossed into Savory's house had burned the home to the ground, so he knew that having enemies wasn't something to take lightly. He moved to the United States and with his U.S.-born second wife, Jody Butterfield, formed the nonprofit Center for Holistic Resource Management in 1984. A separate organization, the Savory Institute, which more directly addresses land restoration and reversing desertification internationally, was founded in 2009 to ensure what Savory considers the most important aspect of his work: using the holistic framework to address the social, economic, and environmental complexity facing governments and international agencies in forming policies, laws, and regulations germane to agriculture.

Virtually by definition, Holistic Management has a lot of moving parts, but Savory's pivotal insight was that grasslands, grazing mammals, and pack-hunting predators evolved together. So if domestic herbivores can be managed such that their behavior mimics that of their wild counterparts, the grasslands—the African savanna or the U.S. prairies and plains, terrain that represents about 45 percent of all land worldwide—will regain the state of wild land: healthy, diverse, and resilient.

For one, the grasses need to be grazed and trampled by the herd effect of bunching animals. Animals eat plants and stimulate their growth during each growing season; the act of grazing and the trampling of animals during the non-growing-season when the aboveground parts of the plants are dead cycles dead plant material back to the surface. This allows sunlight to reach the low-growing parts as well as covering the soil between plants with litter so essential to maintaining the effectiveness of the available rain. The animal urine and dung provide fertilizer as well as the impulse to move on, as an animal will not feed where it has dunged.

Another key is herd behavior. Say a herd of antelope (could be zebra, wildebeest, giraffe, or gazelle; sheep, deer, bison, or caribou) is happily grazing. At some point a predator mob—a pride of lions or a pack of wild dogs; in the United States we'd have wolves—comes on the scene. The antelope stand in a cluster as their main defense against pack-hunting predators, and it's this frequent "bunching" that's so important. The bunching and moving gets repeated across the terrain, leading to the continual defoliation of plants and trampling of litter and soil. This, in turn, ensures both soil cover and the establishment of new plants. The regular movement means that a broad swath of plants get a good nibbling but none are overgrazed. Nor are they over-rested, which would result in accumulated dead plant material that blocks sunlight and hinders new growth. Instead, the herd's trampling—which seems to completely trash the area, with flattened plants and hoof marks everywhere—drops plant residue onto the ground where the microbes can get to work decomposing it. Without this, the dead growth would oxidize, killing grass plants when sunlight can't reach the ground-level growth points that are out of grazing harm. At the same time, the soil underneath the plants would dry out, initiating the slide toward wasteland. Here, the organic litter helps retain moisture for plants to draw upon. The pounding of hooves also pushes seeds onto the ground where they can germinate.

This is the kind of dynamic that had occurred in that trampled spot Savory saw on that South African farm, and why the soil was healthy. Albeit in a small corner of a single plot, the sheep on that farm had acted on the ground in a way that built soil. Savory says the decrease in populations of wild predators in many of the world's grasslands has been hugely detrimental to the state of the land, since it's the cheetahs, hyenas, and wolves of the world that keep the grazers bunched and continually moving. In the absence of predators, however, moving groups of livestock according to a planned schedule can approximate the effect on the land.

One of the tenets of Holistic Management that's left many in the scientific and ranching communities scratching their collective heads is the notion that *more* rather than *fewer* cattle is better for the land. Yet Holistic Planned Grazing typically doubles or even quadruples carrying capacity while improving the land and storing carbon.

The planning process à la Sandhurst military college is not a one-off. Ongoing monitoring ensures that moves planned months in advance are still appropriate given varying seasons and conditions; should anything drastically change, like fire burning over a large area, replanning takes place immediately. It's a continual process of studying the land, adjusting for evolving conditions, and revisiting the plan. One benefit of this model, Savory likes to point out, is that it involves minimal cost; much of the work is in the planning.

Today some ten thousand land managers are practicing Holistic Planned Grazing on more than forty million acres around the world. Interest has spread beyond the ranching community to nonprofits, government agencies, and municipal leaders seeking ways to protect and restore land and watersheds as well as climate change mitigation. The Holistic Management decision-making model can be applied to other processes that benefit from planning, such as governing a town (the mayor of Buena Vista, Colorado, Joel Benson, is a Holistic Management educator and brings this approach to local governance), running a business (see: Buena Vista Roastery, the café Benson runs with his wife, Laurie), and setting personal life goals.

By the way, Savory, who splits his time between Zimbabwe and New Mexico, has always hated cattle. As someone whose heart was with Africa's wild game, he saw domestic cattle as intruders and wanted them off the savanna. After he came to understand their potential role in ecological restoration, however, he decided that he "loved the land, wildlife, rural communities, and humankind more than I hated livestock."

That was the breathless-encapsulated-speed-writing description of Holistic Management. If this were a class, this is the point where I'd pause, glance around the room to make unobtrusive eye contact, and ask, "Are there any questions?"

Now let's zone in on a few points that relate to improving soil:

1. **Disturbance**. In our everyday lexicon the notion of "disturbance" has a negative connotation, so it's counterintuitive to consider it a healing mechanism for land. But land doesn't exist in a vacuum, and what outwardly looks like disruption or turmoil can serve as a biological and ecological stimulant, triggering important processes.

With bare or degrading soil, periodic animal impact creates opportunities for life to find its way in: a loosening of soil that frees up oxygen for microorganisms, and allows a seed to embed in a moist spot. By contrast, without the occasional disturbance the soil surface caulks over so that water can't infiltrate. Upon grazing, grass roots respond by dying back (to be broken down as organic matter, thus keeping carbon in the soil) and shifting plant energy toward producing more leaves and then growing new roots. All these effects promote photosynthesis, biodiversity, and retention of moisture and carbon, thereby nipping the desertification process. Land needs disruption at a certain degree and regular intervals—a rhythm similar to that of game in the wild.

2. **Decay**. When we consider the basic life cycle—birth, growth, death, and decay—that last piece, decay, often gets short shrift. Yet, says Savory, "it is a living process essential to the maintenance of all environments, including grasslands and rangelands." He also notes that each year plant material amounting to billions of tons dies in a compressed few months in seasonal rainfall environments (whether high or low rainfall). Since grass plants don't drop their dead leaves as trees do, these remain as part of the plant even as they wither away. Plant matter that dies during prolonged dry periods doesn't rapidly decay biologically but instead breaks down gradually through chemical and physical processes—oxidation and weathering—which can take a very long time and lay waste to many of the perennial plants that form the basis of healthy grasslands by preventing the full spectrum of light reaching new growing points each season. This does not occur where the atmospheric humidity is high throughout each year and microorganism populations are always high, breaking dead material down rapidly and biologically. In other words, humid areas are friendly to microbes, and biomass breaks down swiftly. Seasonably dry areas, however, need help from herbivores. Their chewing and digesting, their symbiotic relationship with the particular microorganisms that reside in their gut, break down the plant fodder as it moves through the various digestive stations/stomachs and, ultimately, out in a form that fertilizes the soil. Abe Collins refers to a grazing animal's digestion system as a "biological

accelerator," in that it hastens the decaying process, the breaking down of a plant into its constituent nutritional parts. The older, rougher forage that wouldn't sit well in the bovine gut would be crushed by hooves into soil-covering litter, an action that kick-starts the process of decay by other means. In a sense, desertification can be thought of as an affliction of the decaying process: This is where the cycle gets stuck.

3. **Brittleness**. Among Savory's contributions to our understanding of desertification is the concept of brittleness. This term refers to the distribution of humidity throughout the year in an environment. The relative brittleness of a place—a tropical rain forest would represent the non-brittle extreme while a true desert is most brittle—determines the type of management the land needs in order to stay vibrant. Brittleness is usually described according to a 1 to 10 scale, with 1 the most non-brittle and 10 the most brittle. The assigned number can serve as shorthand for how a given area responds to fire, resting, and animal impact.

   Brittleness should not be confused with dryness or amount of rainfall. Land can receive lots of rain and still be brittle. What matters is how that rain is distributed throughout the year. Brittle landscapes have long dry spells, with the year divided into wet and dry periods. At Savory's Zimbabwean ranch, for example, the dry season lasts eight months. My husband marvels at the fact that Johannesburg and London have the same annual rainfall—twenty-four inches—and yet he vividly remembers moving from South Africa where the sun was always shining to England where it always rained. Johannesburg's rain comes in brief, intense bursts whereas London's is doled out steadily but frugally. When I'd asked Peter Donovan about his cross-country trip for the Soil Carbon Challenge, he remarked on the change upon crossing the one hundredth meridian through the Great Plains, how as he traveled east he could observe the switch from a dry environment to one with higher moisture. This was the shift from brittle to non-brittle. Once into non-brittle territory, Donovan told me, the piano in the bus started to go out of tune.

According to Savory, some two-thirds of the world's landmass has seasonal rains and thus falls on the brittle side of the scale. Since it lacks continual moisture to keep insects and microorganisms happy, this land is vulnerable to desertification. Without occasional disturbance and a means to kick-start the process of decay, the land will likely degrade. In brittle landscapes, rest is also damaging. In low-brittle areas—like London, or where I live in Vermont—resting land doesn't cause trouble. The humidity when it's above freezing ensures that organic matter keeps cycling.

The brittleness scale also explains why farmers and ranchers can do everything they're supposed to do and still see the land mysteriously deteriorate. The problem is that the conventional approaches to agriculture that we rely on today were developed in non-brittle environments. Taking a farming model that worked fine in, say, Sweden or Scotland, and applying it to land in southern Africa, Israel, or the American Southwest, has been disastrous to the land's long-term viability. In chapter 2 we saw how quickly land degraded in Australia, where much of the terrain is highly brittle, when it was farmed as if it were an extension of Europe. A brittle environment needs animal impact to help it along: to remove and recycle dead plant growth and to crack hard soil surfaces to allow for the flow of air and water. Likewise, it depends on ruminants' guts to perform functions of decay that moisture would take care of in non-brittle areas.

Bringing Back Land in Africa

The centerpiece of Savory's work and the demonstration site for Holistic Management is the sixty-five-hundred-acre Dimbangombe Ranch near Victoria Falls in northwestern Zimbabwe, home to the Africa Centre for Holistic Management. The Dimbangombe story begins in 1992, when Savory donated land he had purchased in the 1970s to form a nonprofit Holistic Management learning site and social welfare organization to benefit the local population. (A larger parcel of land owned by Savory is now the Kazuma Pan National Park, part of the five-nation Transfrontier Conservation Area.) The ranch is a collaborative project between the Savory Institute and a local nonprofit group; the board of trustees includes all five of the local tribal chiefs. It borders more than a million acres of communally owned land. In this region everything depends on the rains that come between November and March; after that it becomes progressively hotter and drier until the next rainy season. Since both parcels have the same soils and rainfall, the communal lands serve as a convenient "control" to highlight Holistic Management's effect on the landscape.

The name *Dimbangombe* means "the place where the people hid their cattle in the long grass." The local population struggles with poverty, HIV/AIDS and other infectious diseases, the breakup of families and communities as men and young people seek work elsewhere, and the relentless deterioration of their land. In this hot, dry region, "the rains are not what they used to be" is a frequent refrain. But Dimbangombe looks as though it's been smiled upon by the rain gods. It has lush, varied grasses and flowing rivers and streams, graced with reeds and water lilies. Grasses stand waist-high where not long ago ground had been bare. The thriving livestock herds—cattle and goats, some four times the number per acre as neighboring lands, even in drought years—run with their wild counterparts, including sable antelope, buffalo, elephant, waterbuck, reedbuck, kudu, and warthog. Lions, cheetahs, leopards, wild dogs, and hyenas have returned to keep the wild game moving while herdboys (the Southern African equivalent of "cowboy") tend the livestock. They avoid using fencing, which is damaging to wildlife. And the livestock are run in a predator-friendly manner using "lion kraals," or portable panels, so that none of the lions needed to keep wildlife populations healthy have had to be killed.

Thanks to the renewed flow of the Dimbangombe River, now a mile longer than before the implementation of Holistic Management, elephant herds no longer have to travel to the one permanent pool that had been their sole water source. Women who used to walk as much as three miles a day for water are beginning to see the benefits as rivers show the first signs of returning health. Now that people are using livestock for cropland preparation according to their grazing plans, crop harvests are averaging three to five times their past yields. In Savory's words: "Cropland preparation here involves very concentrated trampling, dunging and urinating on the land by tightly bunched animals over a short period of time—only a few days so that trampling does not begin compacting lower soil layers as occurs the longer the trampling.

Farmers then plant the maize and other crops directly into the soil without further treatment—fertilizing, plowing, etc. And the women have no need to physically move manure from home pens to fields."

The adjacent areas have far fewer animals, domestic and wild, and are plagued by dry, barren soils; the plant life is dominated by shrubs, weeds, and annuals as opposed to deep-rooted perennials. In many years, people in the Hwange area, with a population of about 145,000, have depended on international aid for food. ACHM is now working closely with ten communities, as well as promoting Holistic Management education through the Dimbangombe College of Wildlife, Agriculture and Conservation Management and other programs.

Recently, a Zimbabwe government minister visited the ranch. He saw the river and was told that it's now flowing through the dry season higher up the system than it's been known to before. The minister said, "What witchcraft is this? You must have had a lot more rain because how else can water appear where it has not existed in a hundred years?" The head herder, who is illiterate, was able to explain to the minister how increasing the livestock and planning the grazing to mimic nature had caused the water to flow again—as he explained, the hooves of the animals make the rainfall effective again, the way it was in the past.

The transformation has taken time. As Jody Butterfield, co-founder of the Savory Institute and director of its Southern Africa programs, recalls, "Each year things got better and better. Gradually over the years the grass was thickening up and the ground would close in, covered with plants. Then we started noticing, 'oh, the wetlands are expanding along the upper reaches of the river.' We started seeing sedges and reeds growing many yards up from the riverbanks and could now see a huge swath that was becoming wetland. In the past few years especially, it's been quite dramatic."

Butterfield and the others at ACHM were able to watch the land degradation process in reverse as desert was slowly unmade.

Holistic Management is one reason for the transformation. The other, which also has huge implications for reversing desertification, is that Savory, Butterfield, and staff minimized grass fires.

In Africa and many places around the world, fire is the chief means of clearing decaying plant material and promoting fresh growth—functions Savory says grazing animals are uniquely equipped to do better. Controlled fires do play a role in land management in certain ecosystems. But Savory says that in Africa fire is often the default response, to the great detriment of the land and its wildlife. Poachers also may torch the grass to obliterate their tracks, and these fires can get out of hand as well. Grassland and woodland fires can go on for hundreds of miles, turning skies dusky gray as far as can be seen.

"Africa is burning to death, many parts of it," he says. "Over 809 million hectares [nearly two billion acres] of grasslands are burned annually. The reason we're burning them is that there are not enough herbivores to keep the grass alive. This while people continue to believe we have too many cattle and, in particular, too many elephants."

From a desertification standpoint, these uncontrolled, often uncontrollable fires present several problems. They destroy litter and leave exposed soil. This makes any rainfall less effective, which leads to desertification. Rain that soaks into the soil largely evaporates over subsequent days, up to 80 percent. If rain arrives in large storms, most of the water simply runs off as flash flooding. Fires also encourage fire-dependent plant species over the more diverse, soil-enriching grasses that animals eat; kill microbial life; damage air quality; and not only release huge amounts of carbon dioxide, but by destroying plant life also remove potential carbon dioxide sinks. According to French research quoted by Savory, a 1.5-acre fire puts out more, and more damaging, atmospheric pollutants per second than four thousand cars. Many fires burn thousands of acres for several days, making grassland burning a major cause of climate change and desertification. A NASA Fact Sheet identifies Africa as the "fire center" of the planet.

At Dimbangombe, they put in regular firebreaks as a barrier against bushfires. The lion kraals have also turned out to be useful breaks to fire when moved along boundaries.

To return to the government minister's comment, no witchcraft was invoked to bring the rain. Nor did Dimbangombe receive more rain than its neighbors. What's important, Savory says, is not the amount of rain but to create land conditions that enable whatever rain falls to be effective: to keep that rain in the soil where it gives rise to plant growth, biodiversity, the amassing of soil humus, and the recharging of aquifers. In short, the antithesis of desertification.

You may be wondering: What about trees? Since trees hold the soil, store carbon, offer shade and shelter, and provide food and fodder, won't tree planting stop encroaching deserts? Well, Savory says, it depends.

Trees cannot survive in every landscape. "At a conference, an agroforestry technique harvesting water and growing trees was presented as the answer to Niger's [desertification] problems," says Savory. "I asked what percentage of Niger has the 600 mm (24 inches) or above annual rainfall required for the trees being planted. The answer was 11 percent of the land. In Kenya, where they would mandate a certain number of trees per hectare on croplands as the solution to desertification, the cropland areas are 9 percent." In many countries there's but a small portion of arable or irrigable land. After that, he says, "there's the rest that no one wants to talk about." Savory says it's this land (about 90 percent of the land in many countries that is not cropland), over which grass that stores more carbon than trees and supports livestock and provides more soil-covering litter, that "holds the key to our future."

Agroforestry, which involves combining trees and crops together, holds promise as a cropping practice where rainfall is adequate. Still, Savory says, "even here people would achieve more healthy land if livestock and crops were integrated through Holistic Planned Grazing" to build on the synergistic benefits. Similarly, variations like Farmer Managed Natural Regeneration (FMNR), a reforestation technique used in Africa in which trees are sprouted from live tree roots or stumps, can be effective in certain regions, but only when carefully and appropriately applied. Savory is concerned that since tree planting sounds like a good, noncontroversial idea, governments and NGOs might rally around that when the money could be better used to focus on the far more vast areas of grasslands with rainfall too low to provide full soil cover with trees.

While at a conference in Israel, Savory visited the site of an anti-desertification effort in the Negev desert. "They were spending 10,000 Euros per hectare to plant trees on water being channeled off the desert," he says. "They were doing this in the Nabatean civilization over 2,000 years ago right there. It failed then as it is failing now. And while this is being done, the Bedouin pastoralists who could be being taught to use greatly increased sheep numbers to reverse the desertification are suffering cultural genocide due to unnecessary livestock reductions."

Another factor to consider, says Savory, is that "trees, although they play a vital role in ambient carbon cycling to sustain all life, do not store excess carbon from fossil fuels, fires and soil destruction in the way well-managed grasslands do so, safely." Soil, particularly the soils of perennial grasslands, can lock up carbon for long periods of time whereas trees can burn, get chopped down, and die of multiple causes, all of which release carbon dioxide. As we saw in chapter 2, with proper management perennial grasslands can sequester increasing quantities of soil carbon. Says Savory: "Everybody talks about the need to plant trees, but trees can't take in all that carbon, nor do they have the 'pulsing' root systems associated with grazing that effectively move carbon to soil life for centuries."

Not that simply switching from planting trees to sowing *grass* is the answer. "Planting grass cannot reverse desertification because it does not address why the original grass plants died in the first place," says Savory, pointing to the hundreds of thousands of acres of grass plantings in the western United States that were failed attempts to address desertification. The only way, then, is for it to work biologically. Which means focusing on the biological cycles that hinge on the soil—and correcting the practices that initially made the rainfall less effective. To Savory, soil cover is the key: Remove pack-hunting predators and large herbivores stop bunching; annually dying grass plants begin oxidizing, which exposes soil and kills new grasses; rainfall becomes less effective, leading to desertification as a manifestation of biodiversity loss and contributing to climate change. Again, one process as opposed to three distinct problems.

In Burkina Faso, Yacouba Sawadogo, a peasant farmer, has revived a traditional technique to restore lifeless soil with great success. Yacouba, whose mud-hut-to-Capitol-Hill story is featured in the documentary *The Man Who Stopped the Desert*, sought to improve soil on his family farm after years of drought devastated his region. He started in the early 1980s by digging zai pits into dry, capped soil in order to collect rainwater, and expanded them to hold more water. He then placed a bit of manure in each hole to stimulate microbial activity and give his millet, sorghum, and sesame seeds a better chance to grow. Over several years he continued to refine his methods, laying stones in contour patterns in a way that slows flowing water so it has a chance to seep into the ground. He added termites to the manure to speed up microbial decomposition. The termites played roles that earthworms might elsewhere: digesting organic material; aerating and turning the soil; enhancing water infiltration.

Termites, infamous for wreaking havoc on wooden structures, have their ecological niche. Dutch geographer and agroforestry consultant Chris Reij explains: "Termites are a pest risk during extended drought periods, but that risk is reduced in the zai as these pits also harvest and concentrate rainfall and runoff. The termites are vital for improving the structure of the soil and this increases infiltration. Termites can also draw upwards plant nutrients that are too deep to be reached by the crop roots. Their advantages outweigh the disadvantages."

Soon tree saplings began to appear in the zai pits amid the crops. At harvest, Yacouba cuts the millet stalks at twenty inches so that the remaining stalks protect the young trees from browsing by livestock. He nurtured these trees, and today a thirty-seven-acre forest of varied indigenous species grows on his land. Yet there are always some fields reserved for crops to supply grains for his family. While he's created a forest, Yacouba made an effort to share his ideas. Word spread, not just to nearby villagers but to scientists and aid organizations. And eventually the media: In 2007 Mark Dodd, then a cameraman with the BBC, learned of Yacouba from a friend who had been to Burkina Faso on vacation (and said: "There's an interesting man who lived down the road . . ."). Dodd headed down to Yatenga Province to see for himself. "It's a small town in the northwest of the country," Dodd told me on the phone. "People only stop there on their way to Mali. I went to Yacouba's farm and eventually he turned up on his moped with an ax over his shoulder—that's what he uses to dig the zai pits." As the film ends, Yacouba, long the village pariah for his "mad" ideas, is revered throughout the province and, in 2009, travels to Washington, DC, to share his story with policy makers.

Yacouba's model addresses both the need for soil disturbance (digging the zai) and decay (via manure/compost and termites). The area receives more rain than the land Savory typically works in, which meant that the trees could thrive. At the same time, the condition was such that the soil was barren with a hard crust; nothing was growing there. "This soil would not improve even if it were untouched for fifty years," says Reij. "Digging planting pits is among the few ways this soil can be made productive. Digging disturbs the soil and the use of manure combines water and fertility."

Reij has visited Yacouba and was impressed by what he'd been able to create from seriously degraded land. To him, the beauty of the project is that it integrates trees and croplands and relies on the natural generation of trees, as opposed to plantings. The trees decrease the wind speed, he says, which helps protect the crops. The point is that creative common sense combined with an understanding of land function can yield results.

Chapter Four
The Return of Lost Water

The water's still there now
But hidden in the air now
In the clouds it makes a home
Until there's rain to share now.

—from "Water Cycle" by Meish Goldish, to be sung to the tune of
"It's Raining, It's Pouring, The Old Man Is Snoring"

THE WEATHER'S BEEN AWFULLY WEIRD. I write this in southern Vermont in early March, 2012, a day after the temperature climbed above sixty degrees and I wore a T-shirt when I walked the dog up the mountain road. Last night we had terrific winds, the limb-clipping kind, which after dark whipped into a rainstorm punctuated by the occasional crackle of ice and hail. This in a winter during which one January afternoon I changed my boots in an eerily empty Prospect Ski Mountain lodge as the owner and cook stared, drop-jawed, at the Weather Channel: nothing but too-warm weather and rain as far out as they could predict. *Rain in January?* It's been the worst cross-country ski season anyone at Prospect has ever seen. The high school Nordic team—which my son competes on—has been roaming the county, desperate for skiable snow.

Actually, the weather's been weird pretty much since I moved here fifteen years ago. Our first spring there was the tornado that rode 67A through North Bennington and across the Bennington College campus (where we then lived). The sky turned pickle green and I curled up on the couch to wait it out. The minute the winds stopped, a hundred chain saws broke into song, the neighbor-helping-neighbor sound of vigilante road clearing. Last summer brought us Hurricane Irene, which left more than a dozen Vermont towns with no roads in or out for days. From our spot on the mountain I thought it was just a regular old storm, until we got our power back and, via Facebook, shared witness to covered bridges collapsing and cars swimming in parking lots.

Everyone's got similar stories these days. The heat waves are hotter, the hurricanes heavier, the lack of snow more lacking. The extreme and unusual—what weather pros call "hundred-year events," meaning that they come along but once a century—are now extreme yet usual. Meanwhile, our political candidates are asked if they "believe" in climate change, as if it matters what any of us thinks. Unless we willfully choose not to, and many do, we all know that the weather has meandered far from its familiar patterns, that it's gone off kilter. And that this likely has something to do with what we've been doing to the planet.

In news reports about extreme weather events it's become de rigueur for the reporter to ask some highly credentialed authority, in that blithe tone of offhand curiosity we tend to favor in our newscasters, "whether this [fill-in-the-blank: heat wave, rash of tornadoes, relentless wildfire season] is a consequence of global warming." And the expert—no doubt mindful of the job-threatening controversy that could erupt upon connecting too many dots—will say something along the lines of, "Well, of course it is hard to attribute any particular occurrence to global warming but the statistics are showing a clear trend toward intensified storms/warming/instability that is consistent with the model of global climate change." And then, perhaps, we'll hear a vague analysis of the jet stream configuration or ocean currents, all in a matter-of-fact voice, a manner that offers some reassurance that nothing too disturbing is going on and that the experts have it all under control.

I feel strongly that the public should not be protected from knowledge of the consequences, climatic and otherwise, of environmental degradation. At the same time, I have to admit that a direct link between rising carbon dioxide and our increasingly chaotic weather isn't quite satisfying to me. I'm plenty convinced that human activities have altered the ecological conditions that govern climate. And, thanks to that handy Keeling Curve, it's clear that carbon dioxide levels have been rising. But exactly how these intersect doesn't seem clear. What's a reasonable time line for rising carbon dioxide to cause palpable change? Does anyone know? Could excess carbon dioxide, and excess carbon dioxide alone, be making our weather crazy right now? It seems we can have but two choices: to deny that anything is happening at all ("We've always had climate cycles" or, "What's wrong with warmer weather?"), or to attribute all warming and climatic oddities to greenhouse gases, primarily carbon dioxide. Because despite the hugeness of this problem, despite the threat that climate change represents to life as we know it in every corner of the globe, the carbon dioxide narrative remains the only one we have with which to talk about it.

Could there be another way to understand the swings of hot and cold, wet and dry, fluke and flukier, that mark our weather in the here and now? I think there is, and that something important has been missing in the debate (and even our non-debates) about climate change: the role of water in climate regulation.

Now, I didn't just come up with this by myself. So I think it's time I introduced the folks who made water's importance to weather and climate clear enough for me to understand. There is a group of hydrologists and scientists in Slovakia and the Czech Republic who have written a stunning book called *Water for the Recovery of the Climate: A New Water Paradigm*, which I bumped into on Peter Donovan's Soil Carbon Coalition website. Interestingly, the vision articulated by the New Water Paradigm has a close parallel to the Soil Carbon Coalition's project: Just as Donovan seeks to return carbon to the soil, these European thinkers say we need to return *water* to our soils.

Too much water has drained from our land, they say, as a result of deforestation, intensive agriculture, and an expanded built environment. Every field that's turned into a parking lot or industrial park means more water siphoned off into gutters and culverts and less water in the ground. Land that's dried out and lost its soil carbon to oxidation means that rain, unable to soak in, cascades across the surface and into the waste stream. Eventually all that water flows off to the sea. One consequence is that we're wasting perfectly good water at a time when many places are experiencing water shortages. Another is that in letting land-based water slip away, we're losing an important means of keeping our climate in balance. Even if by some political or geophysical miracle we slash our atmospheric carbon dioxide concentration down to pre-industrial equivalents, they argue, we would still contend with the climate havoc wrought by our heedless treatment of the land and of the water that rightfully belongs there.

You see, when people make the connection between water and climate change, they tend to go in just one direction: anticipating how climate change will put added stress on water resources as the planet warms. Scientists, for example, predict reduced snowpack mass, which is troubling since many regions, notably the American West, rely on snow as a reservoir for usable water. Experts also expect that increased drought and wildfires coupled with rising energy demand will tax existing water supplies, particularly as warmer weather will speed up evaporation in lakes and rivers. These are significant concerns. However, scant attention has been given to how the water cycle—or more precisely, the small and large, green and blue water cycles—affect the climate. Or the fact that these dynamics hinge on the soil, as soil moisture plays an important role in maintaining the earth's natural air-conditioning mechanism. This is a crucial topic to explore because, as it turns out, there's a great deal that we can do to support water's heat-regulating potential. And in doing so, we can help regain temperature stability, microclimate by microclimate or possibly on a large scale.

The folks behind the New Water Paradigm, Michal Kravčík, Jan Pokorný, Juraj Kohutiar, and a few others, see themselves as a group of friends as opposed to an organization or advocacy group. Given their backgrounds, these guys were primed to be upstarts. Just months before the fall of communism, Kravčík was forced out of Slovakia's Institute of Hydrology when his research showed that the government-sanctioned water-management approach was ineffective. Kohutiar was a political dissident during the last years of communism, active in the Movement for Civil Liberty and the Catholic Underground. Their ideas are as likely to develop in the course of an evening of chatting over wine as through funded research or professional conferences.

The lead author, Kravčík, is a hydrologist who from 1993 to 2011 ran a nonprofit called People and Water (L'udia a Voda in Slovakian) that provides education and solutions on water and ecosystem management. He was awarded the 1999 Goldman Environmental Prize for the "Blue Alternative," a water-use model that substituted for a large, ecologically destructive dam project at Tichý Potok in the Carpathian Mountains, a plan that cost a fraction of the Slovakian government-supported dam and secured the survival of four 700-year-old villages. Born in Ukraine of Slovakian parents who were among tens of thousands of ethnic Ruthenes sent into exile from Czechoslovakia after World War II, Kravčík loved to paint and was long torn between the technical fields and art. He's always been drawn to the Impressionists, "the way they paint water on the landscape, and show the dynamic atmosphere of water and air."

Pokorný is the outlier in that he's in the Czech Republic—he assures me that the two nationalities don't have problems like in the former Yugoslavia, apart from perhaps drinking too much when they meet. He received a PhD in plant physiology at Prague's Charles University and did academic research on oxygen and photosynthesis. He became interested in evapo-transpiration, the process whereby water moves through and out of a plant in the form of water vapor. "I saw that there's more water circulating through plants than we can see, and that this invisible water cycle is important to the biosphere," he recalls. "Soon I was looking at the world from the point of view of plants, looking at water quality and how plants affect water quality. Water enters the plant as dirty water and goes into the air as distilled water. Plants not only give us oxygen, they also produce for us clean water and function as the perfect air conditioning system."

In the mid-1990s Pokorný met Michal Kravčík. "He showed how much was lost when we drain areas instead of keeping vegetation on the land," he says. Kravčík, meanwhile, learned a great deal from Pokorný: "I hadn't understood the connection between water and solar energy," he says. By combining their two perspectives, they were able to see that the wholesale clearing and draining of land is not only a massive waste of water resources but also interrupts the ecological cycles that lend stability to our climate. Considering that the pace of urbanization is such that around twenty thousand square miles of our planet's surface is encased in concrete or pavement each year, they saw this was not a small matter.

Juraj Kohutiar, who has handled much of the group's writing, is a civil engineer who got to know Kravčík when they were colleagues at the Slovak Academy of Sciences' Institute of Hydrology. They reconnected years later. This was during the chaotic transition period just after the fall of communism, when, Kohutiar says, "the new elite had a need for experienced people. They wanted me to work on security, and when I told them I had no background in this they said, 'At least you had experience on the other side, being persecuted by the police.'" He subsequently became the head of counterintelligence for the Slovak Information Service.

Within the water group, Kohutiar has assumed the mantle of resident skeptic, reining in others' (notably Kravčík's) more sweeping statements when they get carried away by enthusiasm. For example, Kravčík said to me, "I am sure that if we have efforts on a massive scale to keep water on the landscape, after ten years we will have no problem with the climate." Kohutiar's response, when I mentioned this on Skype: a polite, measured distancing from the statement accompanied by a wry, indulgent smile.

They each keep busy with writing and lecturing as well as their respective on-the-ground community ecology and educational efforts: Kohutiar with engineering consulting in Africa; Kravčík, building a drought and climate protection program through the Slovakian government as well as his involvement in numerous international projects; Pokorný, running an applied ecological research nonprofit called Enki, named for the Sumerian "god of fresh water and education, the patron of craftsmen and artists," in addition to research in Africa.

They've also tried to get the international climate leadership attuned to water's role in addressing climate change—to little avail. In 2009, Kravčík served as the group's delegate to the climate talks in Copenhagen, COP15, where he presented the *Kosice Civic Protocol on Water, Vegetation and Climate Change*, a cogent manifesto on the importance of water and plant life in sustaining the biosphere. The document prompted some notice but not enough to make a dent in the conversation. "There was some response from people in the general public, yet no response from the people that we would term decision-makers," says Pokorný. "To the IPCC [Intergovernmental Panel on

Climate Change], we have the atmosphere and in the atmosphere are greenhouse gases. If greenhouse gases go up, it's a warmer climate. It's different when you consider the biosphere and atmosphere and the energy balance between the universe and surface of the earth. There's been a simplification of climate change, a train which goes along, driven by lawyers and business." The CO_2 mafia, as they've begun to regard it.

Solar Heating (and Overheating)

To understand how water mediates warming and cooling, let's take a look at what happens when sunshine hits the ground. If the sun's beams strike bare soil, what you primarily get is "sensible heat," which means it generates heat you can feel. If, however, solar radiation falls on moist soil covered with vegetation (and saturated soil invariably has plant cover), the scenario changes. Rather than producing palpable heat, the solar energy is transformed into "latent heat" held in water vapor. This heat, then, is stored in the gas and released where it's cool (in, say, a forest) or *when* it's cool (in the early morning, at which time it will condense and form dew.) Evaporation consumes heat, and thus has a cooling effect. Condensation releases energy, thus creating heat where it is needed.

"This is the perfect and only air conditioning system on the planet," says Pokorný. "Eco-systems use solar energy for self-organization and cool themselves by exporting entropy to the atmosphere as heat. And the medium for all of this is water." By transpiring, plants act as "valves" that release the heat. Plants have small pores, called stomata, on the undersides of their leaves, which open or close to regulate the release of moisture. A tremendous amount of water flows through plants this way; given sufficient water, some wetland plants can transpire as much as twenty liters of water per square meter. Another way to express it is that the soil "sweats" through the plants, as a means of maintaining a cool temperature on the ground and ensuring that soil doesn't lose essential moisture to evaporation.

"Hot air rises," says Pokorný. "Therefore on bare soil heat is rising—even between crops, such as the bare patches between maize plants. In forests, the shrubs and ferns are cooling the air, so the temperature does not go up."

In terms of temperature fluctuation, "among greenhouse gases, water vapor is the big gorilla," Peter Donovan told me (in a conversation that led me to explore his website). "While carbon dioxide may be the primary driver of global warming, there's more water vapor than other greenhouse gases and it traps a lot more heat." Water vapor makes up between 1 and 4 percent of the atmosphere, whereas carbon dioxide is 0.0383 percent (that's the parts-per-million figure we hear about). Water has a greater capacity to absorb thermal energy than any other known substance. At the same time, water vapor is in constant flux, moving vertically and horizontally through the atmosphere and between forms, shape-shifting from gas to liquid to solid and back again, depending on conditions. While carbon dioxide traps heat, water vapor acts as conveyer of heat, alternately holding and releasing thermal energy as it circulates.

"Big gorilla" indeed. The authors of *Water for the Recovery of the Planet* put it this way: "Water evaporation is the most important agent of energy transformation on Earth." And yet, Pokorný notes, "the cooling process of transpiration is often considered an incidental function of plants, rather than a vehicle to moderate and modulate leaf temperature and surrounding temperatures. It's even cast as a negative, in that the plants are said to 'lose' water this way."

It seems that by twisting ourselves into political and rhetorical knots to agree on greenhouse gas emission yet ignoring what humanity has done to the land, we've been going at this climate thing all wrong—or at least missed some important opportunities. When we rip the vegetation off an expanse of land, we're losing the temperature modulation that those plants provided. We haven't been paying heed to the hidden water cycle, nor giving the collective power of plants the chance to manage our climate for us. To avoid excessive heat, what we want is for transpiration to change sensible heat into latent heat. According to Pokorný, each liter of plant-evaporated water converts 0.7 kWh (kilowatt-hours) of solar energy to latent heat, an amount comparable to the capacity of, say, a large room air conditioner. Bare ground devoid of plant cover retains heat and has little moisture to spare. Radiation from the sun, then, just sits there and the soil proceeds to dry out, oxidize, compact, and lose the capacity to absorb water and sustain microbial life, all of which makes it more likely to remain plantless.

We also want more soil/plant moisture wafting into the atmosphere. This is because water in the troposphere, the air that surrounds us, tames immoderate weather. A dearth of water in the soil means less atmospheric moisture, a situation that sets up the conditions for temperature extremes. In a dry environment, the ground heats up and cools down quickly, and there are significant differences in temperature between day and night, summer and winter, mountain and valley. Think of a desert landscape, where it might hover around 120 degrees F during the day but slip below freezing at night. By contrast, in the damp environment of the equatorial rain forest, the temperature stays a balmy sixty-eight degrees around the clock. "That's because the trees and other plants air-condition the system," says Pokorný. On a dry landscape, up to 60 percent of solar radiation becomes sensible heat, while on watered land as much as 80 percent is transformed to latent heat.

A pattern of increased temperature disparity also invites dire weather events like torrential rains, as well as the opposite, severe drought. Here's why: Dried land and built-over surfaces bombarded with solar radiation become "hot plates," microclimates dominated by sensible heat. Hot air inhibits the process of condensation, so rain is less likely to fall on such areas. Instead, the moist air drifts toward cooler regions, like mountains or forests or points north. The cooler zones, then, receive more rain. Particularly in the summer, these can be heavy rains that cause flooding in nearby, low-lying areas. So while some places are too dry, and stay that way longer, others are positively waterlogged. This perpetuates the temperature differential, which can itself be a trigger for extreme weather systems like tornadoes and hurricanes. As Kravčík explains, "The cold zone and hot zone collide, which creates a turbulent atmosphere."

More frequent flooding also obscures the fact that the land is drying out—and that advancing desertification interferes with its ability to absorb water, so that land can be at once desperately thirsty and inundated. In one of nature's more confounding paradoxes, soil needs water to protect itself from water. Land that's well watered uses rain to nourish plants, moderate temperature, and replenish watersheds and groundwater stores. Parched land can only repel it. The result is not only the loss of water but also topsoil erosion and soil sediment settling in lakes, streams, and rivers.

Kohutiar notes that rainfall measurements have reflected the trend of precipitation moving to cooler regions. In Slovakia, over the last century rainfall has decreased 10 percent on the plains yet gone up 3 percent in the mountains. As for Europe in general over the same time period, rainfall has dropped 20 percent in the Mediterranean while rising 20 percent in Scandinavia. Such changes are often attributed to increasing levels of greenhouse gases, notably carbon dioxide. What's actually going on, however, seems more nuanced.

Then there's all the water lost because we're sluicing it all away. As *Water for the Recovery of the Planet* describes it, this has been our bargain with civilization. Since people have grown crops, they've cleared forests for agricultural land and drained fields for the cultivation of grains. Aside from leading to erosion and declines in soil fertility, this has also created a continual need for irrigation and drainage—which, over time, has depleted groundwater sources and left the soil salty and arid. The mechanization of agriculture sped up this process, which now takes place over huge expanses of land. Research suggests that the pumping of groundwater accounts for 25 percent of sea level rise.

The patterns have played out differently depending on the region. In one geographic/historical note that speaks to their own nations' experience, the authors observe: "The Red revolution in socialist countries collectivized the small fields of small peasant farmers, plowed over boundaries and united plots of land into scores, even hundreds of hectares. Gigantic fields with no natural barriers . . . or protected bands of vegetation limiting surface runoff from the land, were presented as great leaps forward."

The move to towns and cities played, and continues to play, a big part in the erasure of water from our landscapes. While farmers and pastoralists depended on rain, people in urban areas came to see rain as a nuisance, wastewater to be dispensed with swiftly, often in conjunction with sewage waste. Over the last several decades we've waterproofed our cities, coating the world's roads, sidewalks, and roofs of buildings with impregnable materials. We're spared the inconvenience of walking or driving in mud, but we're also, Kravčík et al. write, "draining the environment in which we live. We are causing a long-term drop in groundwater supplies beneath our paved and roofed surfaces" and creating "urban hot island" microclimates, which, with the rise of mega-cities, are merging to become urban hot island macroclimates.

"According to our estimates, each year over 700 billion cubic meters of rainwater vanishes from the continents," Kravčík says. "This is water that in the past had been soaked and saturated in soil, and evaporated in the atmosphere." This massive influx of water to the oceans contributes significantly to sea level rise. He says, "Regarding sea level rise, people are still thinking of ice melt and not about the loss of water from the landscape, the water that flows from the continents to the sea."

The way we use and dispose of water has been drying up our land and created conditions that result in lower total rainfall, but rainfall that arrives in heavy bursts as opposed to showers spread over time in more manageable doses. The combination of the two factors—dried-out land and the draining away of rainwater—means that we're steering the solar power in the wrong direction. The air-conditioning function is no longer operating at peak form; it's as if we had a fan that could potentially offer relief but is mostly blowing hot air. Other consequences of lost water, says Pokorný, include the disappearance of humid zones, which are places of groundwater recharge, and the overpumping of groundwater sources in an attempt to boost agricultural yields—which itself leads to the salination, acidification, and an overall ravaging of the soil.

This has got to change, says Kravčík. "We need to keep rainwater on the land. Yet still we're roofing, asphalting and clearing the landscape—all the while draining rainwater out of the city and bringing spring water in from rural areas." Now that he's built the case, he makes a bold proclamation: "The most urgent challenge of present civilization is to understand that the drying out of landscapes has a much more serious impact on climatic change than an increase of CO_2 in the atmosphere."

Moisture on the Move: The Rain Pump

What we've addressed so far is the small or "closed" water cycle. The small water cycle regulates local climate conditions while the large water cycle, the continual back-and-forth drift of moisture between land and sea, governs climate on a broader geographic scale. We've seen how the small water cycle mediates solar heat by way of evapotranspiration and condensation. In the large water cycle, we get water on a macro level moving in two directions. For one, water is continually flowing from the land to the ocean. By definition, land sits at a higher elevation than the sea. Therefore, thanks to gravity, continental fresh water stock perpetually streams downward into the ocean. Because water can't run uphill, the way continental moisture is replenished is that ocean water rises in the form of vapor. Moist winds transport the water over the landmass, where it ultimately condenses and returns to the earth as precipitation.

While it's accepted that the large water cycle determines global weather patterns, the manifold variants and interplays within the system make it hard to pin down cause and effect. Between forcings and feedbacks, airflows and ice floes, the confounding oscillations of El Niño and La Niña, it's challenging enough for the meteorological and scientific communities to explain climate phenomena as they manifest in weather, let alone take a stab at drawing conclusions about the impact of human activity. Our limited understanding about what causes weather incidents is one reason behind all the I-don't-know-ness when it comes to elucidating our weird weather episodes. What does seem clear is that conventional ways of understanding weather don't quite account for the types of weather anomalies we've been experiencing.

Which brings us to the "biotic pump," a theory that first appeared in the literature in 2007 that pulls some of the pieces together, and in doing so turns many meteorological assumptions upside down.

Let me introduce this concept by posing a question that the biotic pump potentially answers: If precipitation derives from moisture brought to land from the ocean, how does that moisture reach inland areas far away from the ocean? In other words, why doesn't it only rain on the coast?

Answer: It's thanks to *forests*. The high rate of transpiration in wooded areas enriches the atmosphere with water vapor. When moist air ascends, it cools, and water vapor condenses, producing a partial vacuum where condensation has occurred. This creates an air pressure gradient, whereby the forest canopy sucks in moist air from the ocean. This moisture now enters the small water cycle described by the forest and its surrounding region, and brings sustaining rains. The biotic pump is the mechanism by which moisture is transported across land. Forests don't merely grow in wet areas—they create and perpetuate the conditions in which they grow.

This theory, developed by Russian physicists Anastassia Makarieva and Victor Gorshkov, has received little attention in the United States, and many in the scientific community balk at its radical reframing of climate dynamics. That's because the biotic pump model posits that it's the flux of condensation that drives horizontal airflows, not the temperature discrepancy between air masses, as had been assumed. "The same principle allows us to quantitatively explain atmospheric circulation patterns in hurricanes and tornadoes—severe weather patterns accompanied by intense water vapor condensation," Makarieva and Gorshkov say. "While theory of moist atmospheric processes is indeed a commonly recognized 'hole' in climate science, the scientific community does not seem to be well prepared to respond to the challenge by radical paradigm shifts." As was the case with the New Water Paradigm, it's been found that evaporation and condensation, processes that both depend on and have implications for the viability of soil, have been overlooked.

Makarieva and Gorshkov note what happens when large forest areas are cleared: The ocean-to-land winds weaken and the rain-making process stalls. They link the unprecedented heat and drought in Russia over the last few years to accelerated deforestation in western Russia. Australia, which has lost 40 percent of its tree cover over the last two hundred years, has in recent decades seen steep declines in rainfall in most regions while certain areas have been pummeled. The lack of tree cover means there's not a strong enough "pump" to draw moist winds to where it's needed.

Pokorný notes how common these types of anomalies have become: "The disruption of the large water cycle explains why water falls at the wrong time in the wrong amounts. For example, it rains in the desert and washes everything away. The rains come all at once." Not that weather chaos following deforestation is a completely modern phenomenon. In presentations, Pokorný quotes a biography of Christopher Columbus written by his son Ferdinand: "On July 22d [1494], he [Columbus] departed for Jamaica . . . Every afternoon there was a rain squall that lasted for about an hour. The admiral attributes this to the great forests of that land; he knew from experience that formerly this also occurred in the Canary, Madeira, and Azore Islands, but since the removal of forests that once covered those islands they do not have so much mist and rain as before."

One situation that seems to support the biotic pump is the plight of the Mau Forest Complex in Kenya, which has undergone rapid clearing. "Changes of the type that took centuries in Europe—the conversion of virgin soil into agricultural land—happened during one generation in Western Kenya," says Pokorný, who has done research in the region. "People referred to it as a 'water tower,' as it supplies the Rift Valley and Victoria area with water. Over the last 15 years 200,000 hectares [nearly 500,000 acres] were converted to agricultural land. The rivers lost water. In 2009, in the rainy season, August to November, the rain didn't come. In recent years, the rain has been very weak. The lack of water stopped the hydropower station run by the Japanese. There's been a total collapse of life under the catchment." More than half of Kenya's electricity comes from hydropower. In 2009, several thousand families were evicted from the land in an attempt to restore the forest ecology.

Gorshkov and Makarieva have been frustrated by obstacles like delayed responses from journals and the refusal of scientific peers to publicly evaluate their work. However, the biotic pump theory does seem to be gaining acceptance, in part a result of historical research linking deforestation to droughts, as is now considered the case with Mayans and Aztecs. The concept of the biotic pump, if correct, brings a new urgency to the need to conserve forests. Makarieva and Gorshkov write: "It is well-known from the estimates of the fresh water reservoirs on land that if the ocean-to-land transport of moisture stalls, the fresh water will totally disappear on land in just a few years." Therefore, soil moisture and the ability of native vegetation to control its amounts are vital to the integrity of the water cycle on land. So as to maintain the needed pressure gradients, condensation must be intense over forests. This requires significant stores of moisture in soil—to allow for the evapo-transpiration that sparks the process.

If there's no soil moisture, there's no evaporation and no rain. Climate activists have decried the plunder of forests, particularly tropical forests, emphasizing their capacity to store carbon. But now we can perhaps see an even more direct bearing on climate. "The impact of increasing CO_2 concentrations on the greenhouse effect can be completely compensated by a relatively minor change in the hydrological cycle over land," say Makarieva and Gorshkov.

Water, Green and Blue

When we think of continental water in an ecological sense, we generally picture the likes of lakes and rivers, topographic markers that would be depicted in blue on a map. In drawing people's attention to water on the land, Makarieva and Gorshkov and the New Water Paradigm group are making a plea for what has come to be called "green water": the water that moves through the small water cycle via the soil and plants. They say this water has been neglected. "Our legislation protects water in rivers, lakes, and underground stores," says Kohutiar. "As for the water we don't see, we don't care about this water at all. Soil is a huge basin for water but we have no laws protecting it. But this water is more important to us than water in rivers, particularly in terms of maintaining local climates."

In the early 1990s, scientist Malin Falkenmark of the Stockholm International Water Institute articulated the distinction between "blue water" and "green water." Blue water is precipitation that ends up in lakes, rivers, and aquifers, whereas green water is water on land: soil water. While we think of rainwater replenishing reservoirs, in fact 65 percent of water that falls as rain becomes green water. Falkenmark argues that we need to do a better job of managing green water sources—particularly in dryland regions, where blue water pools capture little rain and therefore people are more dependent on water held in soil for drinking and agriculture. While we can't predict the effects of climate change, she says, certain anticipated changes, such as reduced river flow and longer dry periods between rains, add urgency to our turning attention to green water.

According to this understanding of water reserves, maintaining green water stores acts as a barricade against "hot plates," protects against erosion, promotes soil microbial diversity, and helps to build soil carbon. This sets up a positive feedback loop that supports vegetation, as the carbon and water cycles tend to follow each other. Christine Jones writes that for every 1 percent increase in the level of soil carbon, a square meter of soil can store an extra 16.8 liters of water—nearly two buckets' worth. The flip side, she notes, is that a loss of soil carbon means a corresponding loss of land's water-storing capacity, and thus green water.

It's become a truism that future wars will be fought not over oil, but over water. According to the journal *Nature*, four-fifths of the world's population lives in areas where water security is threatened. As we enter an era in which additional water crises are anticipated around the world, attention to green water can help us meet this challenge. When we look at water scarcity, we tend to deal with it as if it's a zero-sum game, as a commodity. Indeed, there are corporations trying to corner the market on it. But when we think of fresh water as a commodity, we're picturing water as static, a thing that can be bottled and stored. In nature, water is continually in motion: shifting form, floating in air currents, and flowing according to the contours of the land. We need to view water, blue and green, as part of the commons, and keep it on the land, where it supports ecological cycles and moderates temperature, rather than allow it to stream off to the sea. This will benefit everyone.

Restoring the Climate Drop by Drop

The New Water Paradigm group wants people not only to understand where we have gone wrong with water, but to turn around and get it right. Part of this involves altering long-held attitudes toward water. Says Pokorný: "We need to change our approach from regarding rainwater as an inconvenience that needs to be removed quickly, to seeing rainwater as an asset to be retained in soil and plants." People also need to let go of a fatalistic attitude toward rain. "Through much of history, people saw the rain as coming from the gods and could not imagine that humankind could have any impact on rainfall," Kohutiar says. "This feeling persists today, that rain either comes or it doesn't regardless of what we do."

People need to manage land with water circulation in mind, says Pokorný. He notes that in much of the world, population is concentrated in cities and people have lost touch with how land is managed on a daily basis: "Decisions about land use are often made via computers from air-conditioned buildings." He adds that farmers, who work closely with the land, are under pressure to produce as much as they can per acre so cannot always take care of the land and environment in the way that, were this an ideal world, they would. "I think we should be aware that farmers manage water for us, because it's not in rivers and lakes where water quality can be enhanced. We can try to store it in rivers but the water we use comes from large surfaces, and large surfaces are managed mostly by farmers."

Each of us can choose to assume responsibility for the water that traverses our path, says Kravčík. He envisions cities in which local groups organize water cooperatives and every home has a rain garden. On the municipal level, he says infrastructure resources need to shift from gutters and gullies to watersheds and catchments. He touts the potential of terraces, contoured barrages, stone canals, rock terraces, log structures, earthen bunds: Vehicles for water conservation are many and varied, and generally low-cost. "If you live in a house [depending on where you are], in one year 100 cubic meters of rain might fall from your roof," he says. "Say your house is 100 years old—that would be 10,000 cubic meters of water lost over a century. Then take millions of houses around the globe and you can easily calculate just how much water is lost from the cycle. We need more water in the small water cycle. It doesn't matter if it's as water or in clouds, as long as it's in the system."

The biggest challenge, he says, seems to be nudging the public away from greenhouse gas myopia. "People keep focusing on the negative, the seemingly impossible task of slowing CO_2 emissions from industry. I tell people, fixing the climate is not about lowering CO_2 but about raising *H_2O* in the atmosphere," says Kravčík. "We have to look at the physical behavior of energy on planet earth. When we do, we see the important role played by water. I will continue to share this information with people around the world, and eventually this will have an effect."

For his part, Pokorný laments that this still feels like an uphill battle. "I'm a scientist by breeding," he says. "There are scientists out there much smarter than me who say, 'I believe you are right, but I can't imagine that everyone else who's focused on CO_2 is wrong.' I grew up under communism. Communism runs on the assumption that if everybody says so, it must be true. It's very easy and comfortable to make a mistake with the majority." He suggests that one reason behind the continual emphasis on greenhouse gases is the ease of measurement. "We look at concentrations of CO_2 and methane in part because we are able to model it. It is not easy to measure and describe the physical processes of ice, water and vapor and their dynamic in the atmosphere, or the processes in the soil that serve to equalize temperature."

The physics and chemistry that underlie water–soil–climate dynamics may be complex, but the prescription that rings through *Water for the Recovery of the Planet* is simple: We need to saturate the small water cycle through conserving rainwater on land. "Many civilizations have done water harvesting," says Pokorný. "Everyone who has a yard or garden can do it. Local governments should be doing this too. If we continue to do what we do now—drain land and remove vegetation—we will desiccate our countries."

Kravčík says, "By retaining the water that we are now sending out to sea, we can change dried-out landscapes to fertile green landscapes again. If land is in good condition, that reserved water will recharge soil and water will infiltrate from underground. The system will produce vegetation, and we'll start to recover the whole ecosystem." Now buoyed by his own optimism, Kravčík's painterly side comes to the fore: "You can think of the sun as yellow and water as blue. Together the sun and water make green, which is nature. This is how we make a green landscape. We prime the small water cycle: evaporation takes water up and condensation brings it down. Every drop of water is key to our recovery." He alerts me to a favorite quote, from King Parakramabahu the Great of Sri Lanka in the twelfth century: "Not a single raindrop should be allowed to flow into the sea without first having been used for the benefit of the people." It is hard not to get swept up in Kravčík's enthusiasm. But whatever the relative climate impacts of CO_2 versus water cycle disruptions turn out to be, it seems clear that land–water dynamics have not been fairly assessed. And that we could be better using our water—and doing so could only benefit our ecosystems and well-being.

As a result of People and Water's efforts in conjunction with governmental programs, people in Slovakia are motivated. "In the last eight months, more than 18,000 small water holdings have been built," says Kravčík. "This creates a lot of jobs for poor people, so there are social and economic benefits too." In 2011, the building of retention structures in communities throughout Slovakia employed seventy-seven hundred people, most of whom had been unemployed.

In south-central Portugal, the Alentejo, an effort to restore water function has been under way since 2007. The area is desertifying, with long rainless periods broken by heavy, damaging downpours—the classic brittle landscape scenario—and rampant fires. The iconic cork oak trees are failing from disease. Tamera, a research and training center for peace and habitat restoration in the town of Colos, has, with the guidance of Austrian farmer and permaculturist Sepp Holzer, redesigned its landscape around water-retaining systems. According to Bernd Walter Mueller, a German national involved in the project, out of one containment basin has emerged New South Lake (otherwise known as "Lake 1"). Here, some ninety-three species of birds have been recorded, including many seen only in water-filled landscapes, and the terraces at the shore have yielded an edible landscape with herbs and newly planted fruit trees. Mueller writes: "Many people who visit Tamera for the first time cannot believe at first that it is anything other than a natural lake."

Recently, Kravčík and several colleagues traveled to South Dakota to consult with members of the Cheyenne River Reservation Tribe on a proposal to apply federal compensation funds to implementing the Blue Alternative on tribal lands. The idea would be to build small dams and weirs and assorted catchments along the Cheyenne's rivulets and meanders to capture rainwater for drinking and allow rain to soak into the soil. Tribal elder Candace Ducheneaux told Indianz.com, an Internet news source from a Native American perspective: "Cheyenne River Reservation's clean water sources have been destroyed through poor water management. Most significantly, the damming of the Missouri River at the Oahe Dam." Largely due to limited water sources, the reservation has had a severe housing shortage and the land is desertifying. In the *Lakota Country Times*, Ducheneaux attributes the high rates of usually rare diseases—"off the charts," she says—to toxins in the water, in part from mine tailings (gold and uranium) and industrial waste. The Cheyenne River is on the route of the proposed Transcanada Keystone XL Pipeline. In spring 2012, members of the Lakota Nation went on a forty-eight-hour hunger strike to oppose the pipeline and its effect on tribal ancestral lands and water sources. Not only does such a pipeline present the risk of spills—the present Keystone XL had fourteen spills on US land in 2010, its first year of operation—but the oil extraction process uses and potentially contaminates drinking water from Canada's boreal forest and poses a threat to land and water sources along its path.

Here, as with all Kravčík's projects, the flow of water is not just a desired outcome, but a tool for ecological change. "We need to work with the elements and the energy we have on the earth," he says. "We have sun, carbon and water. We have dynamic processes of using and storing energy. Now, we cannot change the sun and we cannot change the carbon cycle. But we have a tremendous opportunity to alter the water cycle by returning water to the system."

Old Water Paradigm Versus New Water Paradigm

(Adapted from *Water for the Recovery of the Climate: A New Water Paradigm* by M. Kravčík, J. Pokorný, J. Kohutiar, M. Kováč, E. Tóth)

OLD: The water on land does not influence global warming, which is caused by the growth in the volume of greenhouse gases produced by human activity.

NEW: An important factor in global warming may be the change in the water cycle caused by the drying and subsequent warming of continents through human activity.

OLD: The object of research is the impact of global warming on the water cycle.
NEW: A topic worth researching is the impact that changes in the water cycle have on global warming.

OLD: Urbanization, industrialization, and economic exploitation of a country have minimal impact on the water cycle.
NEW: Urbanization, industrialization, and economic exploitation of a country—affecting more than 40 percent of the world's landmass—have a fundamental impact on the water cycle.

OLD: The impact of humanity on the water cycle is negligible and cannot be reversed by human activity.
NEW: The impact of humanity on the water cycle is at present considerable, and its changes can go in both directions.

OLD: The reason for extreme weather events is global warming.
NEW: The reason for extreme weather events is changes in the water cycle.

OLD: Rising ocean levels are a result of melting glaciers.
NEW: Rising ocean levels are a result of melting glaciers on land, but also of a decrease in soil moisture and groundwater levels, as this water flows to the sea.

OLD: The main source and reserve of fresh water is surface water, in lakes and rivers.
NEW: The main source and reserve of fresh water is groundwater: water in the soil.

OLD: Water is used only once and for one purpose and then is sluiced away.
NEW: Water can be used for many purposes, then purified and recycled.

Chapter Five

Beyond Eat Your Vegetables

> Herbage which appears ideal to the chemist . . . is not necessarily ideal for the cow.
>
> —From *Soil, Grass and Cancer*, by André Voisin, 1959

> It is simpler to cure sick soils than sick people, which shall we choose?
>
> —Dr. Charles Northern, 1936

DAN KITTREDGE'S SPRAWLING FARMHOUSE in central Massachusetts is bursting with vitality. The dining room table is a center of commerce. When Tony and I arrive, Dan is in the midst of a transaction with Linda Fuchs, who runs an organic farm in nearby Brimfield, a town known for its vast antiques shows. Dan's wife, Roshni, is in the kitchen, alternately cooking and typing on a laptop on the counter. Their two children, Anya and Sammy, almost five and almost four respectively, are running around, somewhat hepped up since their dad is finally home after traveling for the better part of several weeks giving workshops on growing nutrient-dense food.

He hands Linda a fifty-pound bag of minerals and a plastic bottle of dark liquid. He explains, talking in a quick-fire, high-energy way that I quickly realize, to my dismay, renders note taking virtually impossible, that this is "rock dust—micronized, plus a bunch of micronized humates and ultra-trace elements from seawater." He turns to the plastic bottle and unscrews the lid. A smell like molasses floats by. "Keep it on the tab," Linda says, as she gathers up the soil amendments.

"Oh, that's right," Dan says. "I still have some ginger coming to me." Linda's farm specializes in yellow ginger, fresh turmeric, and sweet potatoes. Improbable crops for this region, perhaps, but he assures me that they're terrific.

An air of happy bustle fills the cool March day, with children perched on the table, Anya and a friend spinning and pointing in pink-and-black ballet-lesson clothes, Sammy resisting and finally reconciling to a nap, and intriguing but unfamiliar spice smells drifting from the kitchen. We could hear the moos, bahs, and bleats of the animals in the backyard. As we talk, Dan, barefoot and in loose, well-worn shorts, takes an occasional phone call, registering in his head an ongoing tally of calls to return.

But for sheer vibrancy and robustness, nothing could have prepared me for what we find in the greenhouse.

"I'll go to the store and pick up some greens," Roshni says, as she prepares lunch for all of us.

"No, I'll go get some," Dan answers casually, and motions for us to follow. We trudge through the mud—it's been raining lightly on and off—and as he opens the door to the greenhouse we are confronted with salad heaven. The simple plastic-covered hoop house is teeming with gorgeous lettuces of all kinds, each row greener than the last, leaves growing full and upright as if this were high summer rather than winter's tail end. "It's not so good now because the greenhouse isn't heated. If soil is the foundation for a healthy plant, these have been stressed because the soil's been frozen a few times." He pulls a leaf, tastes it, and makes an expression that seems to say, *Not great, but not bad.* I catch Tony looking longingly at the lettuce. "I sell to a high-end restaurant in Cambridge," says Dan. "Every week or so I drive it down. They tell me, 'I'll take what you've got.' People are rabid for fresh greens this time of year."

I laugh and note the absurdity of Roshni volunteering to buy salad at the store when these amazing greens are just twenty steps from the house. "She's a city girl," he says with a shrug. The two met 2004 in Bangalore, where Roshni worked in high tech and Dan was working in the office of food activist Vandana Shiva as India's coordinator for the global anti-GMO campaign.

Our Missing Nutrients

I'd come to see Kittredge, who directs the nonprofit Bionutrient Food Association, to learn more about the connection between soil and health. Early on in my quest to learn about the state of our soil, and why it mattered, I'd started to see references to soil mineral depletion. Across the board, levels of key mineral nutrients—zinc, calcium, manganese, iron, copper—in our food crops have declined by an average of 50 to more than 100 percent over the last century. The USDA National Nutrient Database has calcium and vitamin A in broccoli dropping 54 and 75 percent respectively between 1975 and 2010. Over the same time period, levels of iron and vitamin A have fallen 60 and 40 percent. Minerals play numerous roles in the body, from providing material for bone and tissue to facilitating the transmission of nerve impulses to activating metabolic functions. Minerals are present in the soil. Vitamins, also essential for processes that maintain health and vitality, are synthesized by plants.

According to Graeme Sait, an author-educator on nutrition and agriculture in Australia, some nutritionists estimate that the food we eat today has just 30 percent of the nourishment of what our grandparents ate as children. The major reason, says Sait, is declining soil quality, although the way we process, prepare, and transport food also plays a role. Sait notes, for example, that we're now in a situation where we can buy oranges completely devoid of vitamin C.

I wondered about the correlation between depleted soils and the rising cost of medical care. A 2002 study by the Robert Wood Johnson Foundation found that nearly half of all Americans have a chronic health condition. The Centers for Disease Control tells us that two-thirds of adults in this country are overweight or obese, and that obesity rates in children continue to rise. A significant number of infants, some as young as six months old, are now classified as overweight. Pediatric specialists say this can delay a baby's crawling and walking as well as interfere with motor development. Since excess weight puts children at risk for multiple health problems, we can expect that the demand for medical care will increase in tandem with the obesity rates.

Something is wrong with this picture, and I doubt we can place all the blame on McDonald's burgers and mega-size sugary drinks. Could it be that the food we eat is not truly feeding us—and not just highly processed packaged goods with novel-length ingredients lists, but the vaunted produce section as well? Those of us who can afford it buy organic to avoid chemical residues. But perhaps the bad stuff in food is only part of the problem, the other being that some of the good stuff is missing.

We tend to accept a cultural morality narrative that tells us to eat well and stick to plant-based foods and we'll be rewarded with good health and, barring lapses of willpower, a slender-enough body. But what use is all the dietary advice we get from health experts if the food we feel so virtuous about eating is nutritionally inert? Could our startlingly high obesity rates be a sign not of gluttony, nor even a reflection of the class-coded, deceptively neutral term *lifestyle factors*, but a consequence of inadequate nutrition in conventionally grown food? Might people overeat because their bodies are screaming at them that they're not getting enough minerals and other vital nutrients? Most of this country's food-growing soil has not been treated very well—how could the state of our health not be affected by the state of our soil? Because, ultimately, quality of health depends on the quality of food, and food can only be as good as the soil on which it is grown.

When I started scrolling around on the topic, the first sources I bumped into were agriculturalists, doctors, and doctor-farmers, mostly from the 1930s and '40s. They proved such good, informative company that I spent some time with them.

"To be well-fed is to be healthy." Those are the sage words of William Albrecht (1888–1974), a midwestern agronomist and author whose work inspired Acres USA, an organization and magazine devoted to organic agriculture. He also said, "Food is fabricated soil fertility," asserting that beneath the food–health dynamic stands the viability of the soil.

There's the work of Sir Albert Howard (1873–1947) as well—an English botanist and organic pioneer who said that "the health of soil, plant, animal and man is one and indivisible." He observed that "any weakness or defect in the health of any earlier link in the chain is carried on to the next and succeeding links, until it reaches the last, namely, man." For several decades he was a researcher in the Imperial

Department of Agriculture, first in the West Indies and later in India. While working in these colonial outposts he noted that nothing went to waste, especially waste. He became an ardent proponent of composting, believing that fertile soil relies on the "Rule of Return," the recycling and reuse of organic materials. He expressed concern that "artificial manures" (synthetic, chemical fertilizers) and pest-killing sprays led to a "war in the soil" that compromised the life-sustaining properties of humus-rich earth. "In spite of the fact that we speak of her lavishness," he wrote, "Nature is not really luxurious: she works on very small margins."

In his 1945 book *Farming & Gardening for Health or Disease* (later published as *The Soil and Health*), he noted several experiments demonstrating that people whose food was grown in rich, organic soil enjoyed superior health, including one he learned about from Lady Eve Balfour, author of *The Living Soil* and co-founder of the UK Soil Association. In this experiment, students at a New Zealand grammar school were in dreadful condition when they enrolled—nose and throat troubles, gland troubles, dental caries, incipient gout—but became much more robust after they'd been fed from the school's community garden. Howard quotes the matron of Auckland's Mount Albert Grammar School: "The first thing to be noted during the twelve months following the change-over to garden produce grown from our humus-treated soil was the declining catarrhal condition among the boys . . . There was also a very marked decline in colds and influenza. Colds are now rare and any cases of influenza very mild."

André Voisin, a French biochemist and farmer whose observations about rotational grazing inspired Allan Savory (see chapter 3), made the case that animals are invariably a product of the soil on which they live and feed. In *Soil, Grass and Cancer*, he said that any living cell, plant or animal, is a "biochemical photograph" of its environment, and warned of industrial techniques and treatments that upset the balance of nutrients in the soil and led to deficiencies. He looked not only at minerals but also trace elements, so-called dusts in the soil, that despite minuscule quantities are integral to the health and function of living cells. Since their role is to activate enzymes, they've been described as "catalysts of the catalysts."

I kept digging through websites and found more explorers of the realm of soil and health whose stories shed light on different facets of the topic. A few cameos:

- **Sir Robert McCarrison** left his home in Northern Ireland in 1901 to join the Indian Medical Services. Stationed in the north, he noted the longevity and robustness of the Hunza tribe, who scarcely suffered even colds and indigestion, and attributed this to a diet of nutrient-rich food grown on fertile soils (sprouted legumes, root and leafy green vegetables, fruit, milk products, whole wheat flour cakes, meat in small portions and only occasionally). In the 1920s and '30s he performed experiments on rodents: Rats that ate as the Hunza were healthy like their human counterparts, whereas rats on a typical English lower-class diet—white bread and margarine, sugared tea, tinned meat, boiled veg, and the like—had the usual respiratory and digestive ailments of the day and also developed nervous disorders.
- **Lionel Picton** was a Cheshire physician instrumental in the 1939 *Medical Testament*, a pamphlet that assessed the then-twenty-five-year-old UK National Health Insurance Act and concluded that given the poor quality of the day's food, "the efforts of the doctor resemble those of Sisyphus"—a never-ending battle. Picton was a critic of industrial methods of agriculture and food processing, believing that only whole, untreated foods were nutritious. White bread was a particular target; he was known to throw loaves out the window upon finding one in someone's kitchen. In his 1946 book *Thoughts on Feeding*, he wrote: "Fresh food, grown [on fertile soil] confers upon the animals and men that consume it the powers of resistance to germs whose function it is to prey upon and even eliminate men and animals who are devoid of those powers."
- **Charles Northern**, an Alabama gastroenterologist, came to believe patients' digestive complaints were caused by poor nutrition, the inevitable consequence of poor growing soil. He was ridiculed for his ideas, he told *Cosmopolitan* in 1936, "for up to that time people had paid little attention to food deficiencies and even less to soil deficiencies. Men eminent in medicine denied there was

any such thing as vegetables and fruits that did not contain sufficient minerals for human needs." He presented a paper to the US Senate stating the need to restore soil minerals in order to address the nation's health crisis, and, soon after, left medicine to devote himself to soil restoration. "I gave up medicine because this is a wider and more important work," he said. "Sick soils mean sick plants, sick animals and sick people. I'm really a doctor of sick soils."

- **Carey Reams** (1903–85) was a physician and agronomist raised on a Florida farm, as well as a math prodigy who had a long friendship with Einstein. He practiced nutritional preventive medicine and used nutrition-based treatments on patients with advanced disease—an approach that once landed him in a California jail. (His refusal to join the American Medical Association hardly helped his status vis-à-vis the authorities.) As an agricultural consultant, he helped clients grow nutrient-dense crops free of disease, and as a doctor he opened clinics and created "early warning" diagnostic tools that measured energy levels and that drew on an understanding of soil and crop testing. He said, "All disease is the result of a mineral deficiency." He developed the Reams Biological Theory of Ionization (RBTI), which assesses the electromagnetic/energetic component of life-forms based on his understanding that growth derives from the interaction of positive and negative charges in minerals and soil.

As I perused these investigations into soil and health and mulled over the questions they raised, I felt as if I were in a time warp. In their critique of commercial agriculture and processed food, these people could have been writing about today—and yet I knew that, with bigger and badder agricultural weapons (courtesy of the now $125-billion-plus global agrochemical industry) and more artificial sweeteners (like the ubiquitous high-fructose corn syrup) that pretty up taste, things have gotten many times worse since then. If there's a body of work that makes the case that the caliber of soil helps determine baseline health and it's acknowledged that our soils aren't looking too good, why hasn't soil been brought into discussions of public health? As medical doctors, Picton, Reams, and Northern understood even before the advent of the $200-billion-plus prescription drug industry that throwing pills at illness can be a fool's errand if we don't address the problem at its source: food lacking in key elements due to deficiencies in soil.

While I'd never go so far as to suggest that we can simply toss mineral dust on our fields and wave good-bye to cancer, hypertension, and diabetes, you'd think the possibility of heading off even a fraction of disease-caused human suffering by attending to the foundation of our food chain might be worth looking into. Especially since soil can be brought back into balance with minimal expense; by contrast, it can cost upward of $800 million to bring a single new drug to market. Considering the incalculable human toll of illness and chronic disorders, you'd think we'd be peering under every rock in the hope of finding some answers. And I do mean literally searching under rocks.

I was glad that prior to meeting Dan Kittredge I'd read up on this cadre of mid-twentieth-century thinkers. It gave me a context for his comments, since these people form a part of his intellectual heritage, a bank of knowledge and argument, ideas and writings that he frequently refers back to.

Dan and Linda, of the ginger-turmeric-sweet-potato farm, now wrap up their deal with a brief exchange about bolstering "active microbiology" in soil. Which serves as a segue for our subsequent conversation on soil and health. "Once you get a good living system in your soil, it will build the nutrition you need," Dan begins, having joined me at the wooden table. "By living system I mean bacteria and fungi. There are presumed to be ten million species of soil bacteria, and three million species of soil fungi. Typical cropland has about five thousand species, and we need at least twenty-five thousand for the plants to function anywhere near their potential."

The plants we grow need a wide range of microorganisms, Dan explains, and these in turn need access to a full complement of minerals. For example, of those multitudes of microorganisms, 80 percent are cobalt-dependent. "Cobalt is at the center of vitamin B_{12}, which acts as an enzyme facilitator, central to the production of a whole number of proteins," he says. "Most farmers are not addressing the need for cobalt, or other trace elements, and so from the outset [their plants are] facing diminished potential."

Keeping soil mineralized doesn't have to be costly, he says. "If you understand what your particular soil needs [are], you can generally address any deficiencies with rock dust and seawater. You could use inexpensive local raw materials. In this region, common rocks are granite and basalt. I would choose basalt, which has the broadest spectrum." For those less inclined to DIY soil nutrition, there are mineral products, like those he sells, that are selected, processed, and put in formulas to suit different growing conditions.

Bringing Biology to the Fore

Dan grew up on an organic farm, Many Hands Farm in Barre, Massachusetts, about twenty miles away. "My parents have been running the Massachusetts chapter of NOFA [Northeast Organic Farm Association] since the 1980s," he says, with a hint of pride. "I can say things about organic farms that others can't. Just because you're an organic farmer doesn't mean you have the nutrition right. You're not necessarily getting the biology. Often farms are doing the NPK thing [the nitrogen–phosphorus–potassium fertilizer regimen that's the standard in conventional agriculture] by other means, taking the conventional model and substituting organic materials. But not building a biological system. For me, it comes down to whether you're choosing the chemical model or the biological model. To get high-quality nutrition, it's got to be the biological model."

While this might sound like jargon-y shop talk, Dan is making a powerful statement: He's challenging the approach to growing crops that today dominates nearly all sectors of the agricultural economy, the emphasis on the ratios of the macronutrients nitrogen, phosphorus, and potassium. He's saying that soil fertility, or growability, is not about a list of static components furnished to a plant; rather, it arises through a combination of ongoing, interrelated biological processes in the soil.

Let's take a brief historical detour for some background. While strategies to prod the soil to deliver greater yields are as old as farming itself, the practice was traditionally considered more art than science. In the early nineteenth century most scientists accepted the "humus theory" of plant nutrition: that decaying plants and animal manures, the stuff of organic matter, provided food for living plants, which took what they needed via their roots. Since farmers did well enough by amending with composts and rotating crops (so as not to run out of important nutrients), this seemed to make sense. Minerals, or "salts," remained a bit of a wild card.

This all changed with Justus von Liebig's monograph *Chemistry in Its Application to Agriculture* in 1840. The celebrity chemist of his day, Liebig brought two key concepts to the project of growing food: First, using plant ash he analyzed the composition of vegetation, and identified its chief contents to be carbon, nitrogen, and mineral salts, namely phosphate and potash. Second, drawing on the work of fellow German botanist Carl Sprengel, he promoted the "Law of the Minimum" (mentioned in chapter 2 in relation to nitrogen fertilizer). This maintains that yield is determined not by the totality of available resources, but by the *scarcest* nutrient—the limiting factor. Peter Donovan noted in his talk on the carbon cycle that from a business standpoint Liebig would have seen no franchise in carbon since carbon dioxide is free and plentiful in the air. Nitrogen, too, is plentiful in the air, but not in a form immediately available to plants. He directed his attention to nitrogen, and developed a means of applying nitrogen to plants at the roots via ammonia (NH_3), earning himself the moniker *Father of Fertilizer*.

We now had our first artificial fertilizer. The next step was to generate it on an industrial scale. The breakthrough came in 1915, when German chemists Fritz Haber and Carl Bosch were able to fix atmospheric nitrogen using high-pressure equipment. In another historical game changer, the synthetic production of ammonia also lent itself to the manufacture of armaments. Germany availed itself of this opportunity during World War I, in the form of explosives and chemical warfare; and, later, in World War II, as it enabled the production of Zyklon A and B, poison gases used in the concentration camps—a dark irony since, like most German Jews, all of Haber's family were exiled or faced death in the Holocaust.

The versatility of this process meant that after and between wars there was excess production capacity. So as not to let factories fall idle, chemical fertilizer was aggressively marketed to the public. Writes Albert Howard: "The new process of fixing, i.e. combining, nitrogen from the air had been invented and had been extensively employed in the manufacture of explosives. When peace came, some use had to be found for the huge plants set up and it was obvious to turn them over to the manufacture of sulphate of ammonia for the land. This manure soon began to flood the market." As Howard says, the use of chemical fertilizer "was laid on the farmer almost as a moral duty." Fertilizers pumped up crop yields, but at the expense of other (non-NPK) nutrients that were not replenished. Blanketing the soil with "artificials" could mask such deficiencies—for a while.

By emphasizing the importance of soil biology in lieu of the reductive chemical model, Dan Kittredge is harking back to the wisdom of his philosophical forebears. In a chapter aptly called "The Intrusion of Science," Sir Albert Howard refers to the "present-day . . . failure to realize that the problems of the farm and garden are biological rather than chemical."

While the chemical–biological rift might seem a philosophical or even semantic matter, Lady Eve Balfour alluded to the high stakes inherent in the divide when she outright called chemical treatments a threat to "the living soil." As mentioned, chemical fertilizers have the effect of depleting soil compounds that they don't contain, including soil organic carbon and trace minerals. But since nitrogen fertilizers were relatively cheap (especially since these were often subsidized), the path of least resistance was often to add more chemicals. As time went on, crops grown on heavily treated, mineral-poor soil became susceptible to pests and disease. By the 1940s there were inexpensive pesticides and herbicides on the market. For many farmers, each progression along the chemical path meant more dependency, more expense, and less resilience in the soil and the crops that reside there. Wrote Balfour: "If it is the life in the soil that is its most important property, then obviously we must stop killing it with lethal chemical salts."

John Kempf, a farming consultant based in Middlefield, Ohio, and the source of Dan's products, would agree. Kempf, twenty-four, grew up on a fruit-and-vegetable farm less than an hour from Cleveland. His father was the chemical supplier for the local community, and they were, he says, "very heavy chemical users" on the farm. "When I graduated from school after eighth grade, I was given responsibility for doing all the nutrient and chemical applications. When I was sixteen, I was licensed as a professional pesticide applier."

Around that time, Kempf recalls, "something happened" on the farm. "We noticed degraded soil fertility, declining soil health, increasing weed pressure, and poor soil aggregation, all the while disease and insect pressure continued to escalate. From 2002 to 2004 we had unusual weather conditions, very wet. Our yields were only 30 to 50 percent of what we expected. We looked to chemicals because that's what we knew to do."

It soon became clear that chemicals weren't helping. "We rented some neighboring fields, land we hadn't farmed in the past, and I started seeing anomalies," he says. "We'd have cantaloupe, the same variety, planted at the same time, and on our fields we lost most to powdery mildew. Others on new land showed no sign of it. I began to question: what is the difference between healthy and unhealthy plants? What makes some have natural immunity while others don't?" He sought out books on plant nutrition and physiology, learned from biological consultants, and arrived at a holistic system of plant nutrition. "We applied that in 2006 and went completely chemical free. No herbicides. Disease and insect pressure gradually became less problematic. Our yield and quality began to improve."

Kempf and Dan Kittredge met seven years ago at an Acres USA conference. Kempf has since launched a consulting firm, Advancing Eco-Agriculture, and Dan is a dealer-distributor of his products, including the soil amendments he sold to (or bartered with) Linda. The company manufactures liquid nutritional blends and micronized (meaning broken down into extremely small particles, to ease assimilation by plants) minerals and micronutrient blends. Kempf's approach centers on supporting plants through stages of health so that they form complete proteins. This makes them resistant to typical crop pests, which have simple digestive systems and can only break down amino acids, the building blocks of proteins. The next stage of plant health allows for excess energy to be stored as fats (such as omega-3s and omega-6s) and oils, which adds greater resilience and, as crops, nutrition.

"Most people think that plant disease and insect pests are normal, curses of a Creator," Kempf says. "Disease and harmful insects are not normal in healthy plants. If we have, say, tomato plants, and they begin to get mold growth, that's termed a 'disease' and sprayed with a fungicide. Now, if we have bread on the counter and it has mold, it's no longer fit to be consumed. Yet when a fruit expresses those symptoms we think that's normal and we spray with a fungicide and we eat it. How normal is that?

"When you use an herbicide, pesticide or fungicide—that 'cide' part means 'to kill.' With each application, we impact hundreds of species. The biggest single problem with the agricultural paradigm of the day is the warring mentality. It's us against nature: let's kill all these pests. I'm sorry, nature always bats last. It will always circumvent the inventiveness of our attempts to play God. The biological paradigm works with nature, and this seems to produce high-quality, healthy plants with mineral nutrition derived from the soil and air. When the system works, you get functional immunity in plants that is transferred to people in the form of food. We as farmers are responsible for the health profile of this nation. As farmers we can do more to keep people healthy than all the doctors and hospitals combined. Human health is agricultural."

Tapping into Plant Immunity

As for how most farmers today are faring with their crops, Dan Kittredge says without hesitation that "most are doing a poor job. There's an easy way to tell: Look at the plants. Are they healthy?" He asks if I grow tomatoes, and I nod. "Are they healthy and vigorous at frost? Or are they dying from pests or disease?"

Now we're getting personal. Apparently, tomatoes should be popping fruit until the third frost in the fall. I had no idea. I thought Tony and I were doing well because a few years back we somehow managed to avoid the widespread tomato blight. But plants should keep budding and fruiting until they've run out of something—such as a specific nutrient, or water. "There are limiting factors within the system, and when those limitations begin to show up, you get disease and then plants die."

Dan says healthy plants have within them compounds that keep them healthy: their own immune systems. As it happens, these are compounds that we humans require in order to thrive. "A friend of mine who farms said she was taking a look at her son's multi-vitamin and their soil test—every mineral in the vitamin was on the test. For her this correlation really brought it home."

For example, among the small-print list of vitamin ingredients are the trace elements copper and zinc. Among their many roles, says Dan, these minerals are "critical in building [the] immune system and maintaining its strength and vitality." Another micronutrient, manganese, is "directly correlated to plant reproduction and human reproduction. If you have a garden and there's a deficiency, your tomatoes don't fruit. For humans, a manganese deficiency could lead to a low sperm count or the inability to get pregnant." These minerals, he says, "function along the same pathways in plants as they do in humans."

I was getting it now: We humans are but another set of biological beings, dependent on the same soil-derived substances and processes as those lower down the food chain. At a cellular level, we've got plenty in common with plants.

"When we don't have the basic minerals in our bodies, it's because they're not in our food, because the crops didn't have access to them in the soil," he says. If the plants that feed us, either directly or by way of the livestock that graze on them, are deficient, we'll be deficient—and so many of us turn to supplements to provide what we should be getting from a balanced diet. "Sometimes the minerals are in the soil but locked up, not available to plants. The only way to release the minerals is through the activity of bacteria and fungi. If there are pollutants or agricultural inputs you don't get high levels of soil activity. That's the case with organic farms as well."

I say, "Then what? We have to get food from somewhere." I was getting frustrated. I was hoping for a little simplicity, a clear line drawn between organic and conventional. I'd hate to think we need an advanced degree in horticulture to locate decent food.

Dan says he's creating tools to help people discern quality in produce. This will drive higher quality in the food supply: Once they're able to identify better food, consumers will demand it. He says a "bionutrient meter" is in the works: a near-infrared spectrometer that looks like a flashlight or pointer. You'd aim the device at, say, samples of carrots; the reflected light would allow you to gauge levels and ratios of minerals and nutritional compounds in the respective specimens. "The light that bounces back contains the vibration of the minerals in compounds," he says. "Each plant has a spectral signature—the frequency of the life that's coming out of the product. The objective is to correlate that spectral signature with mineral and compound levels and ratios, and use that information to determine relative quality." The Bionutrient Food Association has started researching carrots, cucumbers, and tomatoes to assemble the data sets and algorithms to make this possible.

What our bodies need and, through our senses, seek out are "plant secondary metabolites," he says, which depend on the presence of trace minerals. "These are tannins, essential oils. That which makes basil smell like basil, leeks smell like leeks."

These compounds are called "secondary" in that they don't play a known role in a plant's primary, necessary-for-life functions such as photosynthesis, growth, reproduction. What do they do? Support immunity, attract pollinators, repel pests, detoxify pollutants, promote tissue repair and resistance to stress, and act as antioxidants and antiviral/-fungal/-bacterial agents. Using John Kempf's model for plant health, they are what plants produce when they reach the third stage of health, after they've achieved the capacity to produce carbohydrates and complete proteins. Embodied in the fruits and vegetables we eat, these substances convey similar benefits to us. So far about a hundred thousand plant secondary metabolites have been identified.

Jerry Brunetti, founder of Agri-Dynamics, a soil and crop consulting firm geared to livestock operations, and a speaker on soil, human, and animal health, calls this conveyance a nutrient "cascade effect." For example, the curcumin in turmeric is an anti-inflammatory for the growing plant and, subsequently, for whoever (or whatever) consumes it. Anthocyanins enhance cell growth and immunity in cherries, elderberries, blueberries, and more. These health benefits of plant secondary metabolites explain why every five minutes another miracle food is discovered and everyone runs out to buy, for example, acai berries and pomegranates.

In 1999, Brunetti was diagnosed with non-Hodgkin's lymphoma and told that without aggressive chemotherapy he might not live more than six months. Based on his understanding of minerals and micronutrients and how soil quality affects nutrition, he devised a holistic healing program for himself. Since the cancer he has attacks the immune system, Brunetti worked systematically to bolster immunity, building his diet around foods like raw grass-fed dairy products (rich in conjugated linoleic acids, or CLAs; the minerals calcium, magnesium, and potassium; vitamins D, A, E, K, and the B vitamins; probiotics; and enzymes), free-range eggs, fermented vegetables and soy products, sprouted grains, and highly pigmented fruits. He avoided sugar and refined carbohydrates (which feed cancer cells and prompt insulin production, which spurs cancer growth), many common vegetable oils (which are prone to oxidize and trigger processes that damage cells), processed dairy and conventionally raised meats, and refined soy products.

His health steadily improved. Now, more than a decade later, he gives talks on topics like "Body and Soil" and "Farm as Farmacy." (Several of Brunetti's lectures are available in audio and video form through Acres USA.) He's the one who originally put me in touch with Dan Kittredge.

Smell and taste are "very useful and powerful indicators" of plant secondary metabolites, Dan says, but this isn't sufficient in an industrial food system marked by manipulation at every step. "Nutrient testing will expose the food supply for what it is. Before industrial agriculture and big tilling, we had massive levels of life in the soil with biologically available minerals. We've basically been using it up. We're at a low level of vitality in our soil and our health. It's all connected. If we're willing to be honest about what soil life does, we can re-create a functioning biological system."

He sets a gray plastic box on the dining room table and takes out two pieces. This, he says, will have to do until we get our state-of-the-art bionutrient meter. The two pieces are a refractometer to measure brix and a metal clamp to draw juice so you can put it in the refractometer. When I look confused, he explains that refractometers, developed in the 1830s, measure the extent to which light bends when it passes through liquid. "German vintners used them to decide which grapes make vinegar and which make wine—the quality of the juice. People mistake brix for sugar, but it's not. Brix tells you about the presence of higher-order compounds, which generally correlates with sugar." By "higher-order compounds," he means what plants produce when there's energy left after the building blocks—carbohydrates, proteins, and lipids. As in plant secondary metabolites. The chart with brix scores of the most common fruits, vegetables, and forage crops was formulated by Carey Reams, and so is often called the Reams Chart.

Dan runs a few quick brix measures. It looks easy enough: You need only squeeze out a few drops on the refractometer's glass lens, and then look for the spot where light meets shadow. The number at the line is the brix. A lettuce leaf from the greenhouse measured 7. (Slightly higher than average. Kittredge shrugs: What do you expect at the end of the winter?) A piece of papaya from the fridge: 14, between average and good for this fruit. "You can get a higher brix when something's been around and dried out some," he says. That's simply because with less water, the nutrients are more concentrated. It's also an indication of why brix alone is not a sufficient measure of nutrient value.

I think: *Uh-oh. This brix could be dangerous in the hands of someone with a high compulsiveness potential.* I pictured vendors at our farmer's market cringing at the sight of me. Much later at home, Tony and I test a red onion from our garden that we'd been storing since late last summer. It scores a 6, average. Then we try a ramp (wild leek), hours after Tony had picked it from the woods. It measures 11. I can't find leeks on the brix chart, but figure it would be in the onion family. Perhaps all it tells me is that ramps have more nutrition than red onions. (The thing with brix is that measurements are compared to similar species, apples to apples as it were.) But still, solidly excellent!

Our brixing is interrupted when Roshni shouts: "Dan! They got out again!"

Several animals—goats, sheep, and a young bull—have gotten out of the fenced area, a problem since the house sits on the edge of the road. Dan leaps up and sprints out in pursuit, bare feet and all. Tony, seeming to think it good form, follows. For a few chaotic minutes Dan coaxes, confronts, and throws chunks of wood at the animals—"this lets me blow off a little steam," he confides to Tony—and manages to herd them back behind the fence, though not without much snorting and braying and complaining on their part.

Dan sits back down as though nothing has happened. If anything, he seems refreshed. He continues: "A good way to make sure you have pests and diseases is to do everything we're told to do. Plenty of organic farms have DDT that's washed down from elsewhere. Also, organically approved pesticides still destroy insect life and throw soil dynamics off balance. Go to some organic farms and you'll see there are no insects. People don't know that.

"But I'm optimistic: Restoring life in the soil and in our food is very doable. We can take areas that can't grow crops, identify the limiting biological factors, and work with that. Stone dust and seawater: That's all you need. No one's going to be able to corner the market on these. Revitalizing growing land is cheap per acre. Look at all the money we're spending per month in Afghanistan." For several years Dan was president of the nonprofit Remineralize the Earth; the Bionutrient Food Association, previously called the Real Food Campaign, is an offshoot of Remineralize the Earth.

I raise the topic of GMOs, expecting a long diatribe. Dan mentions one concern, that because the substance is biochemically unfamiliar (not recognizable to our digestive system's software), this activates the body's immune system. He says, "Essentially, your body tries to kill that food. If your immune system is constantly being stimulated, that's a lot of physiological stress." (I look at GMOs from a soil perspective in chapter 7.) Then he shifts gears. "There's no way to get a high-quality crop with GMOs. The answer to GMOs is metrics. At bottom is the matter of quality. They won't be able to compete in the market.

"The farmers growing GMO crops are seeing their plants die earlier every year. Yields are low. They're supposed to save crops from pests, but they're crashing under their own non-living weight. Let's use that. Let's outcompete, and bring transparency in."

The answer to better food, he says, comes from nurture, not nature, which is the genetics. "We're not getting the genetic potential of most of our crops. Only 6 percent to 8 percent is being realized now. We've been lowballing it for a while now. One tomato seed has the potential to make 150 pounds of tomatoes. We are so used to sickly, puny plants that we forget this. Once we start to do a couple of things right we'll see yields double, triple, quadruple; give us several cuttings when we're used to one. The key is working with nature. When you work with nature, the cost per plant goes down.

"If farmers are growing healthy crops using naturally occurring materials, we won't need fungicides and herbicides. If that's true," he adds as an aside, "the money made by industrial agriculture conglomerates can no longer be used to buy senators. If that's true, the food bill won't be written by Monsanto. All that money from agribusiness takes the knees out of our system."

He again draws a line from chemical additives through sluggish soil to poor-quality food and then to poor health, physical and emotional. "In one study, lab rats had calcium taken out of their diet. They started fighting and killing each other. Then they put calcium back and the rats were cuddly and playful again. We can never understand all the various interactions. Nature is complex, functioning at a level beyond our perception. If basic degenerative diseases go down, the demands for pharmaceuticals are lower. If children have better nutrition, they have a greater capacity to learn. If people have greater intelligence, the better our political debates will be. If we're not using nitrogen-heavy fertilizers, we won't have dead zones in the oceans. We can solve a lot of systemic problems through agriculture."

Dan says that bettering the world through farming became his mission over two trips to India in his early twenties. "As someone raised in Western culture I'd never gotten a good answer to the question why—what are we here for?" he tells me. "I read Eastern philosophy. The basic concepts, of consciousness, love, and transcendence, made sense to me. I went to an ashram in the Himalayas and studied ancient meditation techniques. I got to a point where I could tune into other frequency ranges, like the chakras and prana, and could do hands-on healing. It became a part of reality, like picking up a cucumber is a part of reality. I felt so much power and force and potential in that frequency that I felt I could hurt my body—burn the circuits. My basic insight was that while I may have inherent capacities, my body was not in a good enough state to fully explore them without doing damage. If I wanted to follow this path, it was imperative that I rebuild my body in a more functional form.

"When I came back I reengaged in organic agriculture. I'd been managing my parents' farm, but that was because that's what I knew how to do. I didn't have a personal interest in the process. This time I said, 'I'm an organic farmer, young, and I'm still not in good enough condition. We need to improve our food.' In the Himalayas I'd experienced a moment of grace, a profound state of consciousness when it was clear to me that life's purpose was to seek profound love and compassion in every moment, and that this work was part and parcel of that larger objective."

The best means to improve our food supply, he says, is via free markets. "For me, money talks. When the economic driver is quality, our food will improve. Now the emphasis is on quantity. At this point we don't have the empirical metrics to discern quality."

This is precisely what he hopes to change. "Today our quality metrics are volume, rancidity—meaning, shelf life—and protein level," he says. "No one's looking at the higher-order nutrients." Tools like refractometers and the bionutrient spectrometer can calibrate the presence of complete proteins and other nutrients. Which, he says, tends to tally with crops that are healthy and resilient. And this, in turn, makes growing for quality good business.

"You earn more money if you can grow more than two crops in a season. If you get sick greens, you can only pick once or twice before you get flea beetles." He says that just as people accept poor-quality food because they think that's all there is, farmers have become fatalistic about diseased and pest-ridden crops, regarding them as inescapable. "Pests and diseases are life's cleanup crew. This gets rid of the not-good stuff. If it's food for insects, it's not fit for animal consumption. When you've got disease or insects on crops, to me that's the end of the story. That tells you there are incomplete compounds. I've grown up on a farm and brushed potato beetles off the potato plants. The compounds in those potato leaves were lacking, and perhaps only had simple nitrates rather than a full complement of amino acids. The plants would get beetles because they didn't have access to what they needed."

The same holds true for sugars, he says. Insects can digest simple sugars, but not the complex sugars that characterize a vigorous plant. Insects have their own internal bionutrient meter equivalent, and sense the electromagnetic frequencies of the plants they flit past so as to know which to home in on. When plants lack the mineral array with which to build higher-order nutrients, says Dan, "the result is biological breakdown. Let's hold ourselves to a high standard."

The bionutrient spectrometer currently costs several thousand dollars (the technology is already used in the pharmaceutical industry). Kittredge expects this to drop to the $200 to $300 price point within a few years. "I talked to a regional director of Whole Foods, who expressed interest in the bionutrient meter," says Kittredge. "His take was, 'We'll tell our farmers that if their product isn't up to par they've got two years to make it.' It's enlightened self-interest. The public doesn't know about the amazing decrease in the quality of the food supply in the last seventy years. But where we are now is the bottom of the scale. We can only go up. Let's create a better reality. We're all in this together."

Arden Andersen is an Indiana-based farmer/agricultural consultant and doctor who's a proponent of nutrient density as a rubric for enhancing health, and whose work has influenced both Dan and John Kempf. He's long been frustrated with the medical field's typical nutritional advice—eat a balanced diet—as this ignores minerals and other factors contingent on soil. "The thing . . . to understand is that every organism is dependent on its environment," Dr. Andersen said in an interview with *Organic Connections* magazine. "As we change the environment, we change which organisms will survive or perish. As we change the nutrition in the soil and the dynamics of what's going on in the soil, we can set up the appropriate environment for the beneficial microorganisms to survive, not the disease organisms. They don't like the same environment, particularly as it deals with oxygen; it's a question of an anaerobic (oxygen-deprived) environment versus an aerobic (oxygen-rich) environment. The beneficial organisms are dominantly aerobic organisms. Our pathogens, or harmful organisms, are dominantly anaerobic-loving. The environments that conventional agriculture has set up are predominantly anaerobic environments, so they're most conducive to disease organisms."

Once important factors such as aeration, mineral levels, and microbial activity in the soil are addressed, the nutrition will follow, says Dan. And yield, even if this wasn't the initial goal. "Until you grow a seven-foot-tall eggplant, you won't believe it," he says. "What's possible is better than what we have now. The genetics/yield differential is where we can make things happen." This is the space between the genetic potential of a plant and what that plant tends to give us, an interval that until now has often been filled in with chemicals.

It was midafternoon and I felt I'd taken enough of Dan's time. But one thing before we left: I wanted to try the mineral treatment on our own small vegetable garden. Dan led us to a barn next to the house. "We bought this property as a run-down fixer-upper," he says, as we climb over wooden boards. "I put up all the rafters in here myself." Bags of mineral products are piled high against the walls. He deftly clambers up and tosses two different bags down to Tony, who catches them. I am already mentally arranging the garden, as if merely making the purchase will make our patch bloom with crops. But as Dan says, our harvests can only get better.

Chapter Six
The More the Merrier: Biodiversity Starts in the Soil

> "I hate being a worm!" he screeched, his tiny body trembling. "We're the lowest of the low! Bottom of the food chain! Bird food! Fish bait! What kind of life is this anyway? . . . we never even go to the surface unless the rains flood us out! All we ever do is crawl around in the stupid ground. Oh, and how can I forget? We eat dirt! Dirt for breakfast, dirt for lunch, and dirt for dinner! Dirt, dirt, dirt!"
>
> . . . A strange glint fell across Father Worm's eye. "My boy, I think it's time I tell you a story."
>
> —From *There's a Hair in My Dirt! A Worm's Story*
> by Gary Larson, 1999

> [L]ong before [man] existed the land was in fact regularly ploughed, and still continues to be ploughed by earthworms. It may be doubted whether there are many other animals which have played so important a part in the history of the world, as have these lowly organized creatures.
>
> —From *The Formation of Vegetable Mould, Through the Action of Worms* by Charles Darwin, 1881

Gene Goven's farm in central North Dakota produces diversified grains—sunflower, canola, flax, dried edible beans, field peas, oats, lentils barley, various wheats. He's also in the custom grazing business, carrying about 180 cow–calf pairs this spring. (Custom grazing means hosting someone else's cattle on your land, essentially offering them a twenty-four-hour forage buffet for a fee.)

But what he really farms for is species diversity.

Birds, beetles, butterflies; protozoans, mites, nematodes. Grasses, forbs (flowering non-grass broadleafs), and even scrubby or thorny plants that many farmers would look at and think, *Weed.* While many of the breeds he beckons to his land are buzzing, honking, and flying above the earth, central to Goven's project is the life in the soil. A broad range of prairie grasses and other deep-rooted perennials nourishes microorganisms that cycle nutrients and help build fertile soil. At the same time, the richer and more varied the soil community, the more vigorous and diverse the plants that keep the cattle, native fauna, and pollinators happily sheltered and fed.

"Livestock eat more than just grass," he says. "The native prairie clovers I have are 28 percent protein. Why do I need to plant alfalfa?"

Since Goven began managing diversity with an eye toward soil health, the land's productivity has taken a leap. The grain part of the business bumped up its profitability about 30 percent. As for cattle grazing, in terms of pounds of beef per acre the land was able to generate 3.48 times as much in 2010 compared with 1982.

The sweep of resident wildlife also increased more than threefold. "We've got more songbirds," says Goven. "Some mornings it's deafening outside. Visitors came up from the [nearby] Audubon national wildlife refuge, and in the quarter mile from the shoreline up through the grassy brushy woody draw and up over the hill into a prairie pothole, they counted 112 species of nesting birds." (A "prairie pothole" is a shallow wetland area formed by the Wisconsin glaciation—which, at ten thousand years ago, was North America's most recent glacial advance.) "I've been since told there's few places in the world where there are that many. It's because of the soil. What benefits livestock also benefits wildlife." He says an occasional moose and elk visit, too. The night, he says, "is just flashing with fireflies."

It hasn't always been this way, says Goven. "I bought the land at age twenty. I was thinking, *Oh, I want to be a farmer.* A conventional farmer with all the inputs and equipment. I guess I *wasn't* thinking, after all." He embarked on a grain and grazing operation and after a while, he says, "Things just didn't feel right." He started rotating cattle in 1982, using cross-fencing to section off the animals, and when he heard about Allan Savory he helped arrange for him to come to Bismarck. Back then Savory was hardly a popular guy around the agricultural establishment. "A friend (in a government position) was threatened with the loss of his job if he actively participated in the workshop," Goven recalls. "He took personal days with no pay to do it. He sat in the back, in the gallery, and couldn't say a word, even ask a question."

The week-plus course taught Goven how to orchestrate Holistic Planned Grazing, and, as important, offered a structure for decision making and goal setting. Here we're back to the Sandhurst military college's strategic planning guide, which Savory saw could apply to the North American prairie or the African savanna as well as to warfare. Goven's current production statement, the working document that frames and articulates his intention, is to achieve *sustained profits from crops and livestock, and to have a cultural and aesthetic surrounding all for a higher quality of life.*

"That's my road map," he says. "If we have something defined, it will probably become. Here, the soil and land will tend to become what you've defined. It took me a long time to get there—I'll never get there, it's a journey—but once I had the goal things started hanging together. Now I'm managing diversity for soil health enhancement so that all nutrients are supplied by soil, life and the atmosphere. Livestock and people are also part of the diversity. I learned this from Mother Nature, from observing that in natural prairie."

I had the chance to visit Goven during a trip to the Great Plains to see some of the principles I'd been writing about in action. I drove the seventy-something miles from Bismarck, the state capital, past fields of high wispy green-yellow grasses and small pockets of the aforementioned prairie potholes. It's a quiet landscape, animated by subtlety: a breeze that ripples across vast tracts of open country on into the horizon; the scarlet jolt of red-winged blackbirds dipping in and out of the grasses.

I'm greeted outside by Goven, a man in his late sixties with a brushy mustache and the slightly bulky forward-tilting shoulders of someone who's worked outside for many years, and Faith, an alert and friendly Australian sheepdog. I'd dressed according to the weather report, but wind and clouds had come in and I was cold. We stop for a bit inside, where I meet Goven's wife, Christine, and encounter some exemplars of nature's assorted bounty I hadn't expected to see: several caged and highly coiffed Standard Poodles in various sizes and levels of yippiness. Apparently, Christine breeds, grooms, and shows miniature poodles (her company: Pawfect Poodle). One of them, Ice, is a Canadian champion. She also breeds horses (beautiful horses, the soft noses of which I later have the chance to stroke) and barrel-races (a women's rodeo event in which riders guide horses in a cloverleaf pattern around a set of barrels). Prominently situated in the living room is a cabinet filled with Christine's silver belt-buckle trophies. Originally from Germany, she seems a westerner at heart.

Goven's farm is fifteen hundred acres. "That's small around here," he says. "I used to rent land, too. But if I can build soil down, I don't need more land. We get as much on fifteen hundred acres as I used to get on four thousand or five thousand."

Goven has an earnest, low-key manner and a tendency to punctuate his speech with the word *golly* (which he pronounces *gully*). This gives a deceptive impression of his being somewhat unsophisticated and naive. But he's sharp, enterprising, and plenty innovative. One long winter night, he tells me, he devised a reel contraption to make the process of electric cross-fencing less onerous. "I pictured it in my mind and built it." The invention remains in the public domain, he says, because while an agricultural products firm wasn't interested in buying, he feared the company might steal it. "Gully, if it cost $30,000 to patent it, I'd rather just make it available."

He's also constantly juggling complexity in new ways. For example, when he says he manages for diversity he means this on multiple levels, including chronology: "If I seed a field early this year, I will seed it later next year. That breaks up the weed cycles. I'm changing the timing all the time. It sort of keeps things in chaos. If I graze one pasture on June 1, I won't come back at the same calendar time for ten years. The goal is to create the conditions for deeper rooting [of plants], which then creates conditions for building soil."

He's been monitoring the regrowth response, trying to figure out how much recovery time forage plants need before they can again be grazed without compromising their growth. It depends, he says, on what you're trying to grow. "Everyone says thirty days. I say ninety-plus days. By waiting you get forbs: astragalus, vetches, clovers, and native legumes. If they're conventionally grazed, you lose those. What you want is for cattle to get those plant secondary metabolites."

In other words, for livestock to benefit from the immune-boosting, health-enhancing compounds embodied in the full diversity of plants. He says he doesn't worry about invasive species, such as Canada thistle, a spiky plant often referred to as Lettuce From Hell Thistle and officially designated an "injurious weed"; I can attest to the aptness of those phrases after my own battles with it. "Are weeds a problem or a symptom?" Goven asks. "I'm managing for the natural plant community as a whole rather than managing for or against a specific species." On his land, he tells me, the Canada thistle pretty much stop at the fence line. "Weeds are opportunists. If you have deeper rooting and diversity, the weeds don't have an opportunity" to get established.

An Invitation to Abundance

We go out for a ride around the farm, in a large white pickup truck with Faith alert and upright in the back. Goven scans the gently rolling horizon and says, "To me this is nothing spectacular. Just multiples of little things." He points out a prairie pothole. "These are prime breeding grounds for migrating birds. They dry up during droughts. That's part of nutrient cycling. The area dries up and revegetates. If it's continually wet, it smells like a sewer—that's a nutrient tie-up."

As we drive he points out a blue-winged teal, a cliff swallow, a yellow-headed blackbird. I'm enjoying the sensation of being on a safari, with the windows down, off road on uneven terrain, pausing frequently to take in our surroundings, continually checking for signs of movement. Though it's raining lightly we get out by a grassy hill. At first glance the ground is a reedy monotone, but once you focus in the variation pops out at you. Goven shows me wild onion, wild rose, prairie basket flower, old man's whisker, and wild flax—the one plant he knows of that's found worldwide, he says. "The spring flowers are pretty much done. We need a shot of rain to bring out the early-summer flowers." This meadow hosts more than two hundred species. He kneels down and cups his hand around a patch of grass. "There are six species of broadleaf right here. Diversity working together."

He makes a point of the cultural heritage inherent in this multitude of plant types. "I've been volunteering in Minot with Alzheimer's patients," he says. "Some can't remember their kids, but they'll remember native prairie plants they learned seventy or eighty years back. They can even tell me the poisonous plants and the antidotes for the poison—which are always found one arm's-length away."

Most of Goven's land, including this hill, used to be monoculture. Then in 1990 he started growing oats and field peas, both native, together. "I was getting up to an extra four times the rooting depth versus each alone," he says. "Now I have lentils in with the sunflowers. People said I wouldn't get anything because one would take moisture from the other. That's not happening, and the lentils are pulling nitrogen out of the air for the sunflowers." Conventional wisdom generally leaves him unimpressed. "I've found big bluestem and little bluestem together and they're usually not in the same place," he says. "But plants don't read the books."

For Goven, synergy is a tool for improved land productivity. "I like to get ten to twelve different things together," he says. "Cover crops and crop mixtures are good because they release a variety of different sugars and energy sources, which are used by a range of organisms. This will have positive effects on soil fertility, especially with regard to making nutrients available—and for free! These sugars have a positive effect on growth. When we talk about getting the soil to warm up in the spring it may not be a function of temperature that's important, but the function of biological activity."

Teaming with Diversity

What *about* this invisible, underground activity, the life in the soil that Goven contends is what fuels his business? For one thing, there's an awful lot of it. In the first installment of an Australian podcast series called *Life in a Teaspoon*, Christine Jones notes that one teaspoon of healthy soil contains some six billion living creatures—almost as many organisms as there are people on the planet. She says that soil in the rhizosphere—around the roots of plants—could actually contain *trillions* of microorganisms per gram. The most numerous by far are bacteria, of which there are thousands of varieties, many yet to be named and identified. According to a report published in Tasmania titled, somewhat ominously to my ear, *Soils Alive!*, that single teaspoon of soil we've been peering at could hold up to a billion bacteria. If we translated that to its weight in cattle, it's the equivalent of a mass of more than two metric tonnes of livestock per hectare.

With data like that, it's not surprising that ecologist Jill Clapperton, who now runs Rhizoterra Inc., a soil health consulting firm, on her ranch in western Montana, says the volume of living creatures below ground may well be greater than what's standing on top. North Dakota–based soil microbiologist Kristine Nichols adds that we probably know of less than 1 percent of soil biology. "A recent study found that there are more than 1.6 million species of soil biota and another study found over a million bacteria species which is at least two orders of magnitude greater than previous estimates," she writes.

The inner workings of the soil are often characterized as the soil food web, a concept developed by Elaine Ingham, a soil microbiologist and now chief scientist at the Rodale Institute. The soil food web encompasses the community of organisms that dwell in or engage with the soil and all the energy transfers inherent in those relationships. These living beings interact in numerous ways: eating one another, competing with one another, and sharing or trading resources such as carbon or water. In the crowded and bustling universe of soil there are predators and prey, opportunistic pathogens and defenders, shredders and tunnelers, recyclers of minerals and waste. They are named arthropods (which include insects, spiders, and millipedes), annelids (a group that includes common earthworms), bacteria, fungi, protozoans (single-celled animals with a nucleus—unlike bacteria, which have none), nematodes (really tiny wormlike things), plus yet more obscure—and when viewed under magnification, frankly surreal—arthropod critters like Symphyla (blind and weirdly translucent), Collembola (commonly called springtails), and Diplura (two-pronged bristletails).

For the vitality of the land the balance of all these organisms is important, Elaine Ingham has written, notably the ratio of bacteria and fungi. For example, bacteria are less effective at storing carbon. Therefore, if the proportion of bacteria is too high, it's likely that more of the carbon in the soil organic matter will oxidize as carbon dioxide. In conventional agricultural land, bacteria generally dominate as tilling and chemical use inhibit fungi and alter the microbial population.

Intensively farmed soil is also often compacted, which constrains the flow of air, water, and nutrients. The resulting stress and out-of-kilter microbial scenario create an open invitation to pests and crop diseases. As soil scientist William Albrecht has noted, the line between win–win symbiosis and marauding parasitism turns on minuscule margins.

Thinking about all this microscopic life brings you to this teeny world with layers and hierarchies of tininess. It brings to mind Dr. Seuss's *Horton Hears a Who*, a book I know well because when my son was three, for several months he demanded we read it to him twice a night. I think of all the Whos in Whoville, and imagine multiples of mites and microbes and protozoans and fungi—those imperceptible specks—shouting out to Horton and his cynical jungle fellows, "We are here! We are here! We are here! *We are here!*" It's a through-the-looking-glass alternative reality in which earthworms are considered "megafauna," microorganisms in the topsoil that break down organic matter are called "soil livestock," and microbes that colonize the plant rooting zone move together as "micro-herds." This much activity in what looks like lifeless soil is so hard to fathom that it begs for metaphor.

Here's another analogy that requires a metaphorical leap: John Kempf, the Ohio farmer-consultant we met in the last chapter, talks about how soil microbes act as a plant's *digestive system*. He says, "Soil is to the plant as the rumen is to the cow." While this reads like some mystifying koan that transcends rational thought (just say that again to yourself), Kempf means this: In the same way that a cow's rumen is the first stop for the digestion of forage, soil provides the function of predigesting carbon compounds and other nutrients into forms (soluble amino acids and liquefied carbon) that the plant can assimilate. In this way, he says, the plant is essentially "outsourcing" the task of digestion to the soil.

Christine Jones highlights the similarities between mycorrhizal fungi in the soil and ocean krill, the small crustaceans that are integral to the marine food chain. The fungi, she says, perform a similar function in the soil to that enacted by krill in the oceans, linking photosynthesizers to the rest of the food chain. In soil, the photosynthesizers are green plants, while in the ocean it's phytoplankton, which krill feed on. "Almost everything in the ocean depends on the links between phytoplankton and krill," she says. "Vast tracts of marine life would die without the 'krill bridge'. And so it is with soil. Much of the soil food web depends on mycorrhizal fungi transporting the sun's energy, via green plants and the 'microbial bridge', into the soil ecosystem." She notes that, under a microscope, many soil microorganisms do in fact look a lot like krill.

I've ushered you through this underground micro-tour so as to lead into a discussion of the relationship between biodiversity in the soil and biodiversity above ground. Certainly, soil biodiversity can be seen as a microcosm of the wildly varied life that fills the visible world. But the relationship goes beyond that: The diversity aboveground is *a reflection of* diversity in the soil. If the community of life in the soil is limited in scope, the variety of plant life that springs from that soil will also be limited. Which, in turn, will limit the animal, avian, and insect life that can thrive there. While charismatic threatened species like polar bears and penguins get most of the press, biodiversity actually begins in the soil.

This is crucial to acknowledge because biodiversity loss is an urgent threat to the continued stability of life on earth. We're losing species at a rate of hundreds a year, the result of such factors as habitat destruction, warming temperatures, and overharvesting. It's now being said that we're in the throes of a sixth "mass extinction," the first since the dinosaurs met their demise some sixty-five million years ago. The listing of familiar and beloved creatures either gone or at serious risk is heartbreaking: orangutans, mountain gorillas, jaguars, snow leopards, white rhinos, trumpeter swans. But this is not just a matter of sentimental attachment. You see, we don't even know what we're losing—we're still discovering so many new species of life on earth; who knows what's slipping away before we can label them or the extent to which species are interdependent. Whenever a light goes out on a particular species, the ecological balance is thrown out of joint in a way that can affect food chains, resistance to disease, and the dynamics of any given habitat. Also of concern is the loss of genetic variation within individual species, which is important for remaining resilient to stress, disease, and climate variations. Many scientists see biodiversity loss as potentially more catastrophic than climate change. (Though the two are inextricably linked: Climate change is forcing species to adapt or change habitat, which affects the mix of species in specific ecosystems; biodiversity loss limits an area's resilience to climatic changes.)

Much, then, rides on the backs of microscopic biota that most of us don't even know exist. This is true, says ecologist Steven Apfelbaum, even in ecosystems that don't appear highly diverse, above- or below-ground. He offers the example of peat bogs, with their unique, highly acidic soil chemistry that supports neither microbial nor very high plant diversity. "But what's present are often rare plant and wildlife species that are only found in such unusual settings," he says. "This emphasizes that much of the diversity on the planet is also dependent on unique soil settings that may not in themselves have high diversity but that contribute to overall diversity. This encourages us to think in terms of systems: low-, medium-, and high-diversity soils have been found everywhere on the planet and this mix fosters the diversity of life on earth. Now, many of the more diverse settings have been converted to industrial agriculture or urban landscapes with very low soil microorganism diversity. This in part helps explain the declining diversity of life on earth. When one considers that around 45 million acres of the US are either low or devoid of soil microorganism diversity—broadly distributed, as this is the urban lawn acreage of the USA—it is even clearer why biodiversity on earth is declining."

Our usual strategy for safeguarding biodiversity has been species by species. A species is identified as endangered, the alarm is sounded, and the creature (usually an animal, preferably cute even if in an ugly way) becomes a focus for litigation, a media star, or a political flashpoint—or all three—with the goal of bolstering its population. This is costly and the results have been mixed; while there have been some successes (the bald eagle, giant pandas), the list of species in peril continues to grow.

Could zeroing in on soil health help maintain or even build biodiversity? Apfelbaum says that soil can be a catalyst for restoring ecosystems in a way that promotes a flowering of diversity—and he's seen this happen rapidly (within a few years) even on land degraded by intensive agriculture or mining. He attributes this to certain characteristics of soil microorganisms. First, their mobility. The spores of soil fungi are carried on the wind; microbes may be carried on the feet, fur, or feathers of wildlife. "Think about woodchucks or ground squirrels and the soil particles their digging brings to the surface," he says.

Another is that many soil microorganisms are "cryptobiots," a word that reads like what you might encounter in the next PG13 release in 3-D. Apfelbaum explains: "There are species that can encapsulate themselves and enter a suspended life period there and simply wait in the soil for proper conditions." In other words, they hibernate: But rather than waiting for a change in season, they ride out their time until the setting is more congenial. This adds another layer of resilience to soil systems, a kind of "insurance policy" for tough times. It's worth noting, however, that cryptobiotic behavior is hindered by soil chemical treatments.

Yet one more factor is the redundancy among microfauna and microflora species diversity. "Instead of one big mammal predator like, say, the timber wolf in a land ecosystem, there appears to be dozens of top predator species in soil systems," says Apfelbaum. "This redundancy within the foodweb, and thus the complexity of the food chains, lends resilience, responsiveness, and durability to the restoration of soil system health."

The agricultural industry's penchant for monoculture has curtailed soil biodiversity, but this can be reversed. "We know that [following] the conversion of former diverse native landscapes into agricultural monocultures, particularly row crop corn, soybeans, wheat and other crops typically grown conventionally using industrial agricultural techniques . . . soil condition, quality and health decline," says Apfelbaum. "Studies that have examined these changes have suggested that the diversity of organisms in the soils has declined, or ceased to function as effectively or at all. Some of what we know about this declining condition has been learned by restoring such settings from the row crops back to restored prairies, wetlands and other ecosystems that grew in the same soil prior to the industrial conversion. In such reversion studies, the soil life, condition, health and chemistry balances return, usually after three to five years, ten years in the most disturbed settings."

First, he says, the soil tilth comes back once annual crops are replaced by deep-rooted perennials. Then soil organic matter begins to rebuild, as does soil structure. Studies have shown that after a few years the biological life—microbes, fungi, springtails, the whole menagerie—returns as well.

The domino one-thing-after-another pattern of change is well articulated by Australian climate change activist Tony Lovell in a talk from TEDxDubbo: "If you reduce [soil and plant] biodiversity you reduce biomass (plant cover), which reduces photosynthesis, which reduces carbon uptake and oxygen creation, which disrupts nutrient cycling, which reduces fertility, which reduces infiltration and retention of rainfall, which changes soil moisture, which changes relative humidity, which changes weather, which changes climate." However, a focus on soil biodiversity gets the cycle running in the other direction. Once we've got it set in reverse . . . increasing soil biodiversity leads to increased biomass, which increases photosynthesis, which increases carbon uptake and the manufacture of oxygen, which leads to an accumulation of organic matter, which restores nutrient cycling, and so forth. And ecologically, things start to look a lot better.

Hans Herren, president of the Millennium Institute and winner of the 1995 World Food Prize for using biological methods to stave off the cassava mealybug, saving an estimated twenty million lives in Africa, tells EarthSky.org that while we know a great deal about the genetics of crops like rice and maize, "we know almost nothing about the medium in which they live." The soil research that's done, he says, is "mostly on the physical properties—fertilizers, how minerals move in the soil. That's one thing. But when it comes to soil biology, we know very little. And you know why? Because it's extremely complicated. Now we have molecular tools with which we can differentiate organisms, and see what role they play in the soil, and what do we do when we mistreat our soils." He says it's imperative to study the mix of insects, bacteria, and microorganisms in the soil because "the longer we wait, the more difficult it will be to regenerate some of this system."

Underground Heroes

Let's zero in for a bit on the soil megafauna, in particular that charismatic creature of the dirt: the earthworm. The great naturalist Charles Darwin certainly zeroed in on earthworms, devoting some of his last years to their study; his book *The Formation of Vegetable Mould Through the*

Action of Worms, with Observations on Their Habits (London: John Murray, 1881) came out just a year before his death. He conducted experiments on worms, keeping them in pots in his study. He was interested in their habits, their apparent ubiquity throughout the world, and the fact that they "do not possess eyes, but can distinguish between light and darkness." He determined that while completely deaf, common earthworms are sensitive to vibration. Of his worm investigations he wrote:

> They took not the least notice of the shrill notes from a metal whistle, which was repeatedly sounded near them; nor did they of the deepest and loudest tones of a bassoon. They were indifferent to shouts, if care was taken that the breath did not strike them. When placed on a table close to the keys of a piano, which was played as loudly as possible, they remained perfectly quiet . . . When the pots containing two worms which had remained quite indifferent to the sound of the piano, were placed on this instrument, and the note C in the bass clef was struck, both instantly retreated into their burrows.

Darwin observed evidence of intelligence in the way earthworms use leaves, bits of wool, feathers, and the like to plug up their burrows. Their efficiency was particularly impressive to him: "In many parts of England a weight of more than ten tons of dry earth annually passes through their bodies and is brought to the surface on each acre of land; so that the whole superficial bed of vegetable mould passes through their bodies in the course of every three years." The common earthworm was not native to much of the northern United States, including North Dakota. It was inadvertently brought here by settlers in flower pots. Kristine Nichols says that presumably insects and other soil organisms filled that ecological niche.

My sister-in-law Carin Schwartz is a Master Composter and gives lectures on worms in Scotland near where she and my brother live, for a time in a grant-funded position as regional compost coordinator.

Scotland is not a bad place to be a worm; the Scots are looking out for their welfare. There are a number of state- and nonprofit-supported efforts (Zero Waste Scotland; Scottish Allotments and Gardens Society) to keep worms happily fed—meaning to shift compostable waste out of landfills and into worm boxes. For four years compost bins were government-subsidized.

Carin was drawn to vermiculture after hearing Ron Gilchrist, known to many in Scotland and abroad as "the worm man," give a rousing talk on worms and wormcasts—the end product of worm digestion that's particularly high-nutrient-quality for plant growing. She began to organize community composting and ordered ten kilos (twenty-two pounds) of worms (red wigglers, which are prime composters), which Gilchrist personally transported, several hours by car from Fairlie in North Ayrshire. Explains Carin: "He didn't want to send the worms by mail because he didn't want them distressed." On a recent visit to Scotland I had the chance to visit a few compost-savvy community gardens in picturesque towns like Forres (where my brother's family lives), Bothwell (in the shadow of an ancient castle, and where we were treated to beetroot-and-cheese sandwiches), and Fairlie (a picturesque village on the Firth of Clyde and Gilchrist's home base). The sign welcoming visitors to Organic Growers of Fairlie reads: **PROMOTING GLOBAL WORMING.**

Now for a cameo of another underappreciated creature of the soil (which will conveniently bring us back to North Dakota): the dung beetle. While admittedly not the most glamorous of our native bugs, dung beetles—which come in thousands of variations, and are generally identified as rollers, tunnelers, or dwellers, depending on what they do with animal dung—fill an important ecological niche. The scarab, considered sacred in ancient Egypt, was a dung beetle. The female beetles lay their eggs in a ball of rolled dung. Because of the young scarabs' seemingly spontaneous emergence from the earth, they were associated with transformation and manifestation. The Egyptians also associated the beetle habit of rolling balls of dung far larger than themselves with the daily movement of the sun.

To Gene Goven, dung beetles are a valued part of the diverse life on his farm. "For fly control with the cow pies, many people spray ivermectin or permethrin," he says, referring to broad-spectrum treatments against parasites. "But dung beetles do the job in three days. They're my ultimate no-till drill. In the process of putting dung and urine into the soil, they're also burying seeds with their tunneling. Some seeds have to be scarified [in order to germinate] and so when beetles bring seeds down deeper, that's my seed bank."

Strategically timed grazing bolsters Goven's pest control. "I leapfrog and skip paddocks a minimum of a quarter mile from where livestock had been. I leave the fly larvae and eggs behind. If I have active dung beetles, that's an unfavorable environment for flies. In five days' time I have over 90 percent fly control." However, he notes, "A lot of people think dung beetle are another fly, and use insecticide. If cows stay in the same place, they will keep infecting themselves with parasites. It's better to work with the parasite life cycle and manage where the livestock are and for how long."

Three years ago, he recalls, a customer brought some cows in to graze, and suddenly the dung beetles disappeared. "I said to the guy, you put on Ivomec [a pesticide]," says Goven. "He said, 'That was in March. This is May. The label says it lasts only thirty days.' Those cows were only clean after 140 days. Yet the meat can be eaten after thirty days. Makes you ask, 'What am I eating?' I also lost some of the butterflies. Yet less than 10 percent of cows on the land were treated."

When Goven stopped using insecticide on livestock back in the '80s, some five to seven years later a variety of wildflowers started coming up. He says, "I'm not a scientist. But I observe. If a seed is on the surface of the soil, it's vulnerable to surface predation. Dung beetles are bringing seeds down. We've got cycles and cycles within cycles, and it's all about soil health. Sprays are nonselective [in what they kill]. Why would I want to kill something that's good for the soil?"

Somewhat surprising for an area that's rural and flat and hardly known for social diversity, central North Dakota is a high point for building soil-driven biodiversity. This is a place where one sunny Sunday afternoon in Bismarck I was alone on the capitol's vast parklike lawn doing yoga when a young guy with long hair saw me and turned around his red VW to come over and greet me and say how happy it made him to see someone doing a warrior pose on the green. I asked him a question that had been much on my mind: What do you do in Bismarck if you're different or weird? He shrugged and said, "Do yoga."

But soil-wise, these North Dakotans are getting it down.

And word is getting out. From all over the country and the world, people interested in soil enhancement have made a pilgrimage to the modest offices of the Burleigh County Soil Conservation District on a nondescript street on the north side of Bismarck. I, too, made the trip, and met with district conservationist Jay Fuhrer, a trim native North Dakotan who appears to be guiding spirit of this unusual program (though he continually reminds me that it's a *team*). One program the BCSCD has been experimenting with is mixed cover cropping, which might be described as "applied biodiversity" on the farm. "One thing we've learned about cover crops," he tells me, "is that when you apply cover crop combinations to your cropping system, you can accelerate biological time. You can improve the soil health, feed a balanced diet to soil microorganisms, store carbon, improve water infiltration—all this faster."

Nurturing diverse soil biota is a cornerstone of the effort, says Fuhrer. "What do all organisms need? They need a home, food, and a balanced diet. Natural prairie has probably two hundred different species going on. If we put just corn in there, nature's probably looking at us and going, 'You can do a lot better.'" Marlyn Richter, a third-generation farmer and rancher in nearby Menoken, says that from a soil microorganism point of view, a crop monoculture is akin to junk food: "If there's just wheat or any one thing, it's like us eating doughnuts all day."

Fuhrer says, "if you have low crop diversity, you will run into problems and require inputs, which means more technology. In the meantime, nature's going to try to do a lot of things to heal you: weeds, pests, et cetera." These scourges of farmers, he says, are nature's attempt to fix the land, return it to equilibrium. "Cover crops continue to harvest sunlight and give off root exudates—sugars—that feed soil life. If you harvest plants [without putting new plants in] you're telling the soil biology, 'Sorry, we're not feeding you. But we'll be back in the spring.'"

On our way out the door to visit a farm, Fuhrer pauses by a photo of prairie grass and flowers, noting that this is what I'd be seeing had my visit been early fall or late summer, rather than in the spring when the floral display is less dramatic. "A dozen Argentinian farmers were here a while ago," he said. "They used to do no-till cropping systems with high crop diversity and cover cropping. Then the Chinese offered to contract their soybeans. Now it's just soybeans. They looked at our soil health work and they said, 'This is how our land used to look.'"

This last anecdote and Gene Goven's dung beetle tale point to the challenge of maintaining biodiversity at a time when success is measured strictly by yield and technologies are rolled out before we know the consequences. Including the effects on microbes and insects, corners of the ecosystem that, for the most part, have no one to speak up for them. The often devastating consequences to biodiversity on the farm is something that pains Goven greatly.

"It's hard to find open-pollinated corn anymore," he says. "The other day I had eleven flat tires on a drill on 220 acres for a neighbor. That's because of the GM corn—its tough cornstalks. It's slow to break down, it's not being incorporated into the soil." This reminds him of another recent situation: "I was out with [biologists] Pat and Dick Richardson from Texas. We said, 'What's wrong with that wheat field?' Something was definitely not right. We walked into that field and didn't hear insects. The soil looked dead. It had been sprayed twice with fungicide. That was it—there was no soil life. It killed off the good fungi, too."

The topic of crop spraying comes up again as we head back to the house after touring the land, on our passage by Crooked Lake and serial patches of wild rye, bulrushes, western wallflower, cattails. The rain has stopped and the air is cool, gray, and humid, a static kind of sky with a breezy hint of recent drizzle. Goven stops the truck and gazes out the lowered window. "People here are getting cancer," he says. "They're asking questions. With obesity, there's a theory that 40 percent of the time it's nutrition-related. Our bodies are craving micronutrients.

Glyphosate is a chelator of mineral nutrients." This means that with many herbicides, nutrients like iron, calcium, and zinc are locked up, bound to the glyphosate molecule. "The nutrients are tied up and not available in the food so people keep eating. Is glyphosate also tying up micronutrients in the soil? We've been led to believe that it dissipates. But in New Zealand, research found it has a twenty-two-year half-life. Bees, too, are having problems. I'm wondering whether it's from the Bt factor [the toxin *Bacillus thuringiensis*, bred into GMO crops to resist pests] in GMO crops. It's shown that GMO corn is a contributor to bee colony collapse. With GMOs even the pollen in the air is affected."

Glyphosate is the active ingredient in the herbicide Roundup, and the most-often-used herbicide in the United States. Some estimate that more than two hundred million pounds are applied annually in the US. (The next chapter will address this in greater depth.) As for the New Zealand study, I couldn't find that reference and neither could Goven. He says he regretted not printing it out when he found it because such information often inexplicably disappears. "Where does that information go? Why does it disappear? A twenty-two-year half-life means that in forty-four years there's still 25 percent remaining. And more would have built up in the meantime."

I don't get the impression that Goven has planned to talk about this with me. Clearly the effects of chemical inputs on nature weigh on his mind. I have the sense that Goven sees himself as affable, and this topic triggers anger and disillusionment at a pitch of intensity that challenges this comfortable identity. It's hard to be genial and laid-back when you're watching the land, wildlife, and vocation you love take a beating from the industry intended to serve them.

"The major Corn Belt areas are having a decline in bushels per acre," he continues. "Two years ago in South Dakota, the yield of sunflowers was half of what was expected. Now they have micronutrient sprays so they can be foliar applied. 'Have a problem?'" he asks rhetorically. "'Easy—I call the chemical companies to ask what I can do.' The plant is falling over so they breed [firmness] back into it. It's a vicious cycle. When does it end? When is the point of diminishing returns?

"I project: The problems from Roundup and ivermectin will ultimately be more devastating than DDT. Something else to be aware of:

There's a lot of off-label mixing. Maybe five to seven things or more. How do you keep chemicals separate? If they start linking together, those are new compounds [being created]. Twenty miles from here, there are no invertebrates. This is in an Audubon preserve adjacent to fields. The maximum fine for misapplication of farming chemicals is $200. If it's windy or there's an inversion, the most that can happen is that the sprayer gets a fine. I lost a half mile worth of butterflies. The neighbors noticed, too. Less than twenty minutes after I made a call to a regional EPA director, an executive at Monsanto was notified of what was going on. And the next morning the state pesticide coordinator for the state agriculture department of North Dakota called me and said, 'What hornet's nest have you stirred up?'" Two months later, this coordinator was told to take early retirement or his position would be terminated. Goven then made the effort to document the losses on his fields due to spray drift. When he went to follow up on his query, he was informed that his file had been purged.

Relying on chemicals, ignoring warning signs, and silencing anyone questioning their usefulness: That's one route we can take. And this is the road we'll be riding on if we continue to focus on production rather than the processes that underlie all the goodies we get from nature. The agrochemical road is lonely (if you're cheered by birdsong), bumpy, and beset by one crisis after another. But we don't have to go that way. There's an alternative course, which I glimpsed one cool spring day in rural North Dakota: a path lined by prairie grasses and flowers, graced by butterflies and skippers, lit by fireflies. Here the diversity of living things functions synergistically, as opposed to operating in isolation, species by species, in a kind of biological zero-sum game. Organisms we can't see are accorded the same respect as those that are big and flashy or promise to give a nice immediate kick to the economy. Soil is viewed not as an inert granular medium for growing things, but as a hub for valuable activities, interactions and exchanges.

Think we can make that turn? The forces of expediency are aligned against it, but there are some hopeful signs. First, look at how through tending to the soil, land such as Gene Goven's farm can be returned to its biologically diverse splendor.

There's also the fact that more of Goven's peers are beginning to think the same way. He recalls the isolation he originally felt upon applying Holistic Management and questioning conventional views on farming. "Back then, I had to go a long time before I could find someone to rub elbows with." Now that his neighbors are venturing into cover cropping and attending seminars, he says, today "I can rub elbows with someone across the road."

Chapter Seven
The Soil Grab

> Remember that when we talk about agriculture and food production, we are talking about a complex and interrelated system and it is simply not possible to single out just one objective, like maximising production, without also ensuring that the system which delivers those increased yields meets society's other needs.
>
> —HRH the Prince of Wales, in his speech "The Future of Food" at Georgetown University, May, 4, 2011

> We abuse land because we regard it as a commodity belonging to us.
>
> —Aldo Leopold, *A Sand County Almanac*, 1949

When I finally get hold of Amie Bandy, she is in Columbia, Missouri, having just taken a picture of a cornfield that was, in her words, "completely dead." "Here's the deal," she tells me. "Things are cooking." It's the second week of July, a couple of weeks into a brutal heat wave that has hit the midwestern states on top of an unusually dry spring and early summer. With every passing rainless day farmers have been revising their yield estimates downward while Wall Street has been watching the commodity price edge upward. The fate of the corn crop "dominates every conversation," says Bandy, and as heat records fall only to bow to the next extreme, there's talk of a new Dust Bowl. On July 11, 2012, the USDA declared one-thousand-plus counties in twenty-six states natural-disaster areas. The nation's biggest crop, corn, fuels our economy—from packaged foods and sweeteners to animal feed to energy (biofuel) stock—so a price bump would likely ripple throughout the whole system.

"I honestly thought it would have rained by now," says Bandy, a crop consultant based in eastern Iowa. "If I can use this to educate clients, if I can explain why their soil temperatures are 130 degrees, then it will be worth something. The only time we ever get to fix things is when they're really broken. Now crops are dying in a different way from how they did before. This is because of our choices. I'm not sure they're bad choices. I'm just saying there are choices. We've been doing things one way for so long, I think people are ready to consider something different."

Corn-wise, the client she's come to counsel will not be totally devastated, she says, though "a week ago they called me to see if they should bale it for silage [stored, fermented fodder for animals]. Now they understand: My first job is to get them to change. I want them to change bad enough that I'm willing to teach them. To cool the soil down. To find a way that under drought conditions the farm is resilient. We've got to get smarter than the dirt."

She photographed a neighboring, or "across the fence," cornfield so as to have a comparison. This corn, she says, is "all brown, cream-colored. Instead of being green it looks like standing straw." Though she hasn't talked to the farmer, she's guessing he or she applied ammonia (a source of nitrogen) and maybe potassium chloride (potash), which she says would explain the excess heat.

"Outside right now it's ninety-one degrees," she says. "The soil temperature is 113 to 139 degrees. Inside the plant itself it's 103 to 109 degrees, root zone to crown. The ear is where you get your seed. Anything over eighty-six to ninety-two degrees, it's cooking in there. It's getting too hot. Getting too much ammonia, at the expense of the crop." Air temperature matters a great deal—if it exceeds ninety-five degrees during the small window of time when corn is pollinating, the kernels won't grow well—but soil temperature is something farmers have more control over. One way to cool the soil, she says, is with a cover crop. Or with a different fertilizer so that rather than ammonia nitrate they would use ammonia sulfite, which would be more conducive to soil microorganisms. "Forgive the comparison, but microbes are like union workers: They don't work if it's colder than fifty-five or higher than eighty-five. What do you think my microbes are doing now? They went on vacation. Soil is a thermostat. If it's at 135 degrees, it's a broken thermostat."

A big problem Bandy sees is that much of the agricultural community has "bought into the industry's mind-set. People aren't looking at the biochemistry. They've been doing what they've been told to do.

We've always assumed that our university teachers could tell us what to do, that corporations knew the answers. Corporations offer one-size-fits-all solutions. Otherwise they wouldn't be able to blanket-sell." By uncritically buying in, she says, "farmers are accidentally creating their own problems. I'm helping them see what they've been doing on their farm."

Bandy is not against commercially sold agricultural inputs per se. "I came out of this system," she says. She grew up on a farm in Indiana. "My grandparents farmed. In those days if you were a boy you automatically worked on a farm. I left high school and I'm thinking: If I'd been a boy I could have been driving the tractor now." But she wasn't and so she went on to college, completing a BS in soil and crop science at Purdue University. Her consulting business is called "The Crop Advisor" and she regularly speaks at workshops and works with clients throughout the Midwest.

What concerns her is that the way inputs are applied may mean trouble in the soil—sometimes aggravating the very problems a farmer was trying to resolve. She explains: "Glyphosate chelates out [binds in a way that interferes with the nutrient being taken up by the plant] five different micronutrients. But it's not just glyphosate, it's all of them. Like LibertyLink [a genetic engineering product from Bayer CropScience designed to withstand Bayer herbicides]. You have to understand that you're taking out a specific set of elements from the microbial community. You either replace those micronutrients or eventually you're going to be deficient."

Different cover crops pull up different minerals, making them available in the soil and stimulating a given set of microorganisms. For example, oats pulls copper, and wheat draws up silica. "Say you put on a cover crop and still use herbicide—there's not a good way of evaluating the result," she says. "If I want to pull copper up and I use herbicides that destroy copper-promoting plants, am I ahead?"

Her concern was sparked by a combination of problems she was encountering in the field and research that revealed unintended impacts of herbicides, in particular the work of Don Huber, her former professor at Purdue. "I began to understand that what herbicides do in the plant tissue, they're doing the same in the soil tissue," she says. She says she's also started looking at research outside the United States, such as in New Zealand and Poland, that calls into question the assumption that herbicides are benign in the soil.

"First you've got to change the ideology," she says. "Until you can show people that they're losing profits, you can't get them to change." This current crisis may prove the ultimate teachable moment. "I see people getting smarter with some of their inputs."

The drought and heat wave have Bandy crisscrossing the Midwest right now as farmers try to salvage this crop season. "We will have a very challenging year when it comes to corn," she says. "You may think that with high prices these farmers have got it made. But $8 to $10 for a bushel of corn does not keep a farmer farming. They've also got to feed their animals." On the road she's encountering some pretty unhappy farmers. "They don't like it when they look at corn that's dying."

Bandy was measured in her comments about agribusiness inputs and I understand that. I also understand that plant–soil interaction is exceedingly complex; it's difficult to make generalizations. And yet much depends on living relationships in the soil—the biochemical choreography of animal, vegetable, and mineral—certainly to the degree that it's worth being vigilant to make sure that we don't mess it up.

From our conversations, two things Bandy said got my journalistic antennae twitching. First, this comment: *I began to understand that what herbicides do in the plant tissue, they're doing the same in the soil tissue.* Second, that it was primarily research *outside the United States* that presents evidence of deleterious effects commonly used chemicals have on the soil. Why aren't such papers available here? I thought of Gene Goven, who'd looked online for a New Zealand article that reported a longer-than-claimed half-life for glyphosate in the soil, only to find it had vanished. Of the Monsanto executive calling the EPA official within twenty minutes of Goven's expressing concern over an herbicide's toxicity to wildlife and the official who fielded this complaint losing his job soon after. Of a trade journalist I'd heard of who was told in no uncertain terms by higher-ups that he was not to investigate the environmental consequences of agrochemicals. Of research grants being pulled when it became clear results would point to untoward side effects of chemical treatments, thus effectively dead-ending those studies; with only, say, three years of data, these wouldn't get a look from peer-reviewed journals, as five years is considered the minimum standard. But who would fund such research, especially given that much university research (or at the very least the buildings in which it takes place or the professors who lend their name to it) is bankrolled by corporations with a stake in keeping untoward conclusions out of the public eye?

The use of chemical fertilizers and other treatments has increased worldwide as large tracts of agricultural land are turned over to commodity mono-crops. And yet the science addressing the potential impact of these chemicals on the future viability of the land has been locked in secrecy. Here in the United States, we're even kept in the dark as to whether the food we buy contains genetically modified substances, which would necessarily have been treated with herbicides. (The message to those of us who'd like to know what we're eating and feeding our kids is, "Don't worry your pretty little heads over it.")

The rise of biotech in agriculture has led to a sweeping disenfranchisement of citizens and farmers regarding self-determination over what they eat and grow. Consider that farmers are sued when genetically manipulated seeds they don't even want sprout on their soil. (The pretext: This is "patent infringement" and thus the company is owed royalties. No matter that the designer seeds have contaminated the crop.) Or the draconian contracts in which farmers must promise not to save seeds, which has been a fundamental right of growers since the advent of agriculture. Or that tucked into the 2012 Farm Bill was a clause indemnifying Monsanto if the use of their products happens to mess up someone's land. Or corporations pressing their biotech wares on farmers in developing countries, trapping them in dependence and debt. India, pursued as a pilot market for many products (notably Monsanto Bt cotton), has seen an estimated 250,000 farmer suicides over the last fifteen years. Or that large chemical companies like Monsanto, DuPont, Bayer, and Dow, committed to promoting GMOs and synthetic treatments, now control more than half of the global seed market.

To put this as plainly as possible, what we're dealing with is a corporate appropriation of the world's soils. Which is occurring on two levels: the known and potential effects that chemical/biotech products have on the soil, and the question of who owns the land and therefore controls the soil and what it yields. This in the context of diminishing resources, including projected widespread food and water shortages, when all we've got is a capitalist system through which to allocate them. Something that in itself is problematic, since free-market capitalism as currently construed promotes the goals of the corporations (market domination and profit) even when it violates the wishes of ordinary people and their desire for sovereignty over the food they eat and the crops they cultivate and feed to livestock.

These are pretty broad statements, so let's loop back and zero in on one example of how the use of a product plays out in the soil. With that in mind, I now introduce . . . Roundup.

Chemical Experiments

In the world of today's agricultural inputs—the stuff that douses our cropland—Roundup rules. Developed and introduced by Monsanto in the 1970s, Roundup is the trade name for glyphosate, an extremely effective broad-spectrum herbicide, meaning that it kills pretty much any green, growing thing unless it's specifically resistant to it, either by design ("Roundup Ready" or other genetically modified crops) or by happenstance (newly evolving "superweeds" that have managed to outsmart the herbicide by developing resistance). The product was quickly embraced by the agricultural industry because of its wide range of applicability and ease of use. It was also touted as being "environmentally friendly" in that its scope and potency meant that other weed killers could be omitted, resulting in less overall herbicide applied to land; and in that it allowed farmers to minimize tillage, thus diminishing the risk of soil erosion. The active ingredient, glyphosate, was reported to have low toxicity to humans and wildlife and to break down rapidly in soil so that its ecological effects were negligible. The EPA continues to classify Roundup as low-toxicity, despite numerous studies that have linked glyphosate itself or inert ingredients in Roundup to health problems, including birth defects.

As we'll see, claims that Roundup is benign to soil life are debatable as well.

Roundup quickly became the world's best-selling herbicide around the globe and a huge cash machine for Monsanto. This proved pivotal for the company's shift in orientation away from industrial chemicals and toward agricultural biotech. Between Monsanto's dumping of mercury and polychlorinated biphenyls (PCBs) and use of the disfiguring Vietnam War defoliant Agent Orange, the chemical sector was generating damning and costly lawsuits over environmental contamination and human harm. Plus, the chemical industry had matured; market growth was elsewhere. Through various corporate merging and shedding and amassing of divisions with different names and hair-splittingly distinct legal statuses, the Monsanto that brought us saccharin (its first product, introduced in 1901 and subsequently associated with numerous health side effects) and dioxins (a hugely toxic by-product of commercial chemical production) has morphed into the new Monsanto—which focuses on agriculture.

This fresh branding as a "life sciences" company allows for crisp photos and video clips of cornstalks bending to the breeze, smiling farmers posing with their children, picturesque sunsets and sunrises. The copy on the corporate website's homepage reads: *Producing More. Conserving More. Improving Lives. That's sustainable agriculture. And that's what Monsanto is all about.* It's all very sleek and reassuring and high-minded. Like Roundup, the company's glyphosate-based and other herbicides all have Wild West names like Maverick, Lasso, Lariat, Buccaneer, and Ramrod—a list that, to me, seems could just as well double for a line of condoms or, say, a cult series of S-and-M flicks. The theme, I suppose, is all upbeat, can-do, don't-mess-with-me.

Monsanto's exclusive patent for glyphosate expired in 2000, which brought many companies, including price-slashing Chinese manufacturers, to the game. Which presented a business dilemma: namely, how to keep Monsanto's one-stop herbicide gravy train chugging along. Global demand for glyphosate kept growing (more than 20 percent a year through most of the decade that started in 2000, now exceeding $5 billion in annual sales), but by 2008 China had become the world's leading manufacturer. What was a helpless major US corporation to do?

Let's travel back to before the patent expiration to see how the company chose to maintain its market advantage. Here's a clip from a 1997 *Mother Jones* article by Mark Arax and Jeanne Brokaw detailing how Monsanto staked its fortunes on the Roundup brand:

> Monsanto's U.S. patent on Roundup runs out in three years, and if the company is to keep its dominant market position beyond the year 2000, it needs a new spin. Enter Roundup Ready soybeans and Roundup Ready cotton, seeds genetically manipulated so that they can survive direct applications of Roundup. Farmers who once confined their use of the weed killer to the borders of their planting area can now douse entire fields with Roundup instead of using an expensive array of sprays that each target just one or two weeds. "It expands the Roundup market," says Gary Barton, a Monsanto spokesman.

The "catch," the authors continue, is that "farmers using Roundup Ready seeds can only use Roundup, because any other broad-spectrum herbicide will kill their crops. So, with every Roundup Ready seed sale, Monsanto sells a season's worth of its weed killer as well." Then there's the "service" aspect of the Roundup Ready package, which features those contracts barring farmers from selling or even saving seeds for future seasons.

Around 2010 Monsanto "repositioned" its Roundup herbicide business in response to "fundamental structural changes in the marketplace." This included lowering the price (so as to better compete with Chinese generics), emphasizing a "simplified brand strategy" (bundling the herbicide with seeds genetically modified to withstand it), and offering new solutions to "address the need for a simple weed resistance package . . . [with] complementary chemicals" (in other words: *Sure, some weeds may flourish despite our poison—but don't worry, we've got stronger stuff for you.*) The Roundup crop system also launched Monsanto on a seed-company-buying spree, so that it is now the world's dominant vendor of agricultural seed. As North Dakota farmer Gene Goven put it, after decades of inputs, more and more the fallback response among farmers is, "'Have a problem? Easy—I call the chemical companies to ask what I can do.'" This works very well . . . for the chemical companies.

Meanwhile, around the globe farming operations are being consolidated and switched to commodity monocultures, and management is falling to a generation groomed to use chemical inputs as traditional know-how fades into the past.

I haven't even mentioned "terminator" technology: seeds engineered so that the plant produces sterile seeds or no seeds at all. Farmers are then forced to purchase new seeds every year. It's thought that pollen from plants with the terminator trait could infect neighboring crop fields, rendering those plants infertile. Globally, more than a billion people depend on small, often marginal farm holdings for their food. What could possibly go wrong here?

So basically we're in the midst of a dubious biotech experiment with huge consequences for the world's soil and all the plant and animal life that depends on it (read: life on earth), and what drives this highly risky enterprise is . . . *market share.* Meaning employees in a windowless conference room, armed with flip-boards and graphs, discussing strategic opportunity in the scarily aseptic language of corporate marketing. If that's not the ultimate twisted commentary on capitalism, I don't know what is. Yet this is what happens when corporations bear responsibility to their shareholders and essentially to no one else (and the regulators blink).

Forgive me. I tend to get worked up about a company that's got its tentacles in much of the world's food chain and is only angling for more, to the point where by the mere fact of eating every person alive becomes a wholly owned subsidiary of the company [cue trademark sign]. Which to some extent we already are: One German study of city residents found significant concentrations of glyphosate—five to twenty times the limit for drinking water—in every urine sample. And Germany, mind you, is a country that has banned GMOs, so there's got to be less residual glyphosate in the food system than in a country like, well, *ours.*

All right, time for me to stop hyperventilating and instead sit back and bring in some cooler heads. My task is to describe how commonly used biotech tools affect the soil, not to enumerate every questionable act that Monsanto and its cousins have ever carried out. (Of course on that score I've hardly begun. Fortunately, plenty of very sharp people are already on the case. See, for example, *The World According to Monsanto*; gmwatch.org; Organic Consumers Association; Food Democracy Now!; and many others.)

Biotech and the Soil

Let's now bring in Robert Kremer, PhD, a microbiologist with the USDA's Agricultural Research Service and an adjunct professor of soil science at the University of Missouri's School of Natural Resources. For the last two decades Kremer's primary research area has been "the relationship among soil properties, plant growth, and soil microorganisms as influenced by land management practices," which led to investigations of commonly used agricultural chemicals, chiefly glyphosate. His findings suggest that the use of glyphosate—currently applied to hundreds of millions of acres of the world's crop-growing land—has multiple harmful effects related to plant nutrient uptake and soil microbial life.

With glyphosate and other synthetic chemicals and pest-management controls typical of modern, conventional agriculture "several things can happen," Kremer says. "The assumption has been that they will disperse, that the chemicals will be degraded by the microbial communities that are in the soil. After several years of looking at this we're seeing what we like to call 'non-targeted effects,' we're finding that some of these chemicals will shift different species in the soil. Some may be suppressed and others will be enhanced. Often the beneficials are suppressed and the less desirable ones benefit. While the unfavorable organisms are always in the environment, they are usually kept in control through relationships with other organisms."

This creates imbalances in the soil community, which can then alter other soil-based processes, he says. He notes, for example, microorganisms that produce mycotoxins—naturally occurring chemical substances exuded by certain mold fungi—might come to dominate.

One problematic form of mycotoxin is aflatoxin, a substance that if ingested is known to cause illnesses, including cancer, in animals and humans. (Aflatoxin contamination was the cause of frequent pet food recalls in recent years.) Crops such as corn are susceptible to aflatoxin contamination, particularly in hot, dry years (like this one). Disturbed soil communities can also lead to plant root diseases, says Kremer: "It's just a matter of something that is causing the pathogens to be dominant. This disrupts some of the functional activities of the microbial life in the soil, such as decomposition, which is very important for carbon and nitrogen cycling and the nutrient cycle."

He mentions research from Ohio demonstrating that a single application of one pound of active ingredient per acre can provoke an increase in fusarium (a group of fungi in the soil, several of which are toxic and cause crop blight) in just one year. "No one knows how many years it will take after the use of this chemical has stopped for soil to be restored to its original microbial diversity," he says.

While a company's internal product testing may look at one or a limited number of chemical applications, research such as Kremer's offers a picture of what could happen over a period of years. Not only is glyphosate widely used, he says, but it's often applied annually on the same fields, which means it can build up in the soil. "Years ago when we used to use chemicals at several pounds per acre we'd do so to control a pest that was foreseen to be a problem at sixty days, but we did it before planting," says Kremer. Under these circumstances, the herbicide was degrading before it even met the pest, a phenomenon known as "enhanced degradation," in which microorganisms develop the ability to degrade a chemical they've encountered and use it as a food source. "Now we have chemicals that are more resistant to degradation, and are staying in the soil longer. Even though glyphosate is used at a low rate, there are reports that residual chemical can be detected in the soil a year later. It's beginning to build up in the soil, [even if] not necessarily at high levels." He adds that some researchers are finding glyphosate being transported via surface water from adjacent fields. "We don't know what those effects are yet."

Kremer grew up on a farm in mid-Missouri and has been involved in agriculture all his life. He has worked in the US Department of

Agriculture for several decades, and speaking out on the problematic effects of a widely used chemical that's been officially or unofficially sanctioned by the agency he serves is not something he'd be prone to do lightly. He recalls: "When all this biotechnology came about, we all assumed that these chemicals were effective for weed management and there were no other effects. The soil wasn't really considered. Other than the belief that they would be adsorbed completely and thus immobilized. [*Adsorb* means for components of a liquid or gas to adhere to a surface.] That's what everybody just took for granted. So nobody was really looking at the soil. I got interested [in glyphosate], seeing such widespread use and as a microbiologist said, 'Let's see what's happening.' The chemical is applied to foliage. Its effect is systemic—it's translocated through the plant and eventually to the root zone, where it attacks the biology. When it was originally released for nonselective killing of vegetation, we knew that when the plant dies [the residual chemical] would stimulate some [soil microbial] species. But nobody thought about that with transgenic crops. We've found that glyphosate stimulates these pathogens, even if the plant is resistant to the chemical." Of, for example, herbicide-resistant soybeans, he says: "They're stimulating potential pathogens the whole time they're growing in the field."

Kremer observes that glyphosate's effect on soil communities has received little attention. "Roundup-resistant weeds: that's taking all the attention," he says. Perhaps, as I said to him, *Here Come the Superweeds* makes for a catchier headline than *Soil Microbes Run Amok Underground.* In any event, the industry answer for superweeds is more agrochemicals: mixtures of weed killers called "herbicide cocktails." One recommended treatment is 2,4-D, an active ingredient in Agent Orange. Dow AgroSciences is moving ahead with 2,4-D-resistant corn and soybeans that will also withstand glyphosate—crops, then, that can be drenched with two chemicals. Kremer says, "We used to have to mix two or three herbicides. Originally, Roundup Ready crops were supposed to make it so easy since you used just one chemical. Now we're coming full circle, where we have to use as many if not more."

Next I talked to plant pathologist Don Huber, professor emeritus at Purdue University (where Amie Bandy was his student) and a retired colonel, who for more than forty years has done research on biological pathogens, both natural and man-made (germ warfare compounds, say). In spring 2011 he unintentionally sparked a stir when a letter he'd written privately to Secretary of Agriculture Tom Vilsack was leaked. This communication was to alert Vilsack that scientists had come upon a new pathogen associated with plant diseases and reproductive problems (infertility, miscarriages, stillbirths) in cattle, pigs, chickens, and horses and which was found in crops genetically modified to tolerate glyphosate. In an interview with Food Democracy Now!, Huber said this previously unidentified organism can kill a fertilized egg in twenty-four to forty-eight hours. Through his letter, Huber urged Vilsack to pursue more research before GMO alfalfa, the country's main forage crop, is approved and enters the food supply. Numerous scientists, including Purdue colleagues, refuted Huber's statements and claimed he had jumped the gun by, well, calling for caution. Meanwhile, the USDA approved GMO alfalfa, which is now sold under Monsanto's Genuity "trait master brand."

I recently spoke with Huber, whose understated, methodical manner contrasted with the hysterical fear-mongerer image that some, including Monsanto's PR minions, would use to portray him.

"Let's start by going back to the very basics," he says. "We need to recognize that farming is a management program for a system. In that management process, sometimes we forget that there are four major components: the plant, the physical environment of the soil, the very dynamic component of soil microorganisms, and your pests. When we think we have a silver bullet, we forget the interaction[s] among those four components that are so critical to success—to whether we have a successful crop, a nutritious crop, disease or no disease. Any time we make changes in agriculture we change the interaction of those four components. In the same way, one gene operates with all the components. We can't just look at one gene and say it's only doing one thing. We don't have enough genes for all the processes that take place."

Glyphosate, Huber says, affects all four sectors. The chemical works by inhibiting the plant enzyme EPSPS, which is essential to the building of protein. The idea is that the targeted weed basically languishes for lack of protein and stops growing. However, he says, the notion that

EPSPS is all that glyphosate interacts with is "a tremendous fallacy." He refers to Kremer's research on glyphosate and soil biology. Inhibiting that enzyme "predispose[s] the plant to soil-borne fungi that then kill the plant," he says. "You can't kill a plant with glyphosate—not at an herbicidal rate—in sterile soil. It only stumps the plant until it recovers the nutrients that had been tied up. If disease organisms are present, the plant is killed because it has no defenses." Michael McNeill, an agronomist and crop consultant based in Iowa, has said that spraying glyphosate on a plant is "like giving it AIDS."

While glyphosate's big selling point has been ease of use—that farmers would no longer need a variety of weed killers, since glyphosate hit everything—Huber says this has been more than outweighed by an increased need for fungicides. "The mode of action meant that plants would be more sensitive to fungi and other diseases, even the Roundup Ready plants. We have epidemics of fungal and bacterial diseases destroying a lot of plants."

Why have growers and food manufacturers rushed into the Roundup program? One reason, says Huber, is simplicity: "We always look for silver bullets." Then there's business. "Glyphosate is the first billion-dollar pesticide," he says. "We've never had a pesticide that's brought in so much money to a corporate entity." The chemicals have become so intrinsic to our food production that it can be hard to disentangle causes and effects. "We have to realize that what we're seeing today in our own health, animal health and crop production is not normal. It's just that we've been seeing it long enough that we think it is normal. Young scientists think all these diseases and pests are just things we have to live with. That's not true."

He suggests connections among plant disease, animal morbidity, and human health problems: "Look at what's happening with autism. With chronic disease and infertility. All of those have always existed but there's been a 600-fold increase within the last fifteen years. That's not normal." This raises the question of what has changed over this time period, he says. And one change has been the increased use of agrochemicals and the appearance of GM crops (genetically modified food entered the marketplace in 1996). "Glyphosate is so excessively used and abused that it impacts every aspect" of our food chain, he says. "It's even patented as an antibiotic to kill off digestive microorganisms. I don't find it surprising that there are allergy responses and intestinal concerns. It's a very intense biocide for those beneficial organisms."

A crop bred for glyphosate resistance gets plenty of the chemical. As a result, glyphosate is ingested by whatever animal eats it. Chemical on the plant also moves down through the roots and into the soil. The more is used, the more it builds up in the soil system, and the stronger the effects. "In plant nutrition the direct effect is the chelation," says Huber, referring to the binding of nutrients to other elements in the soil so that they're inaccessible to plants. Huber's research found that levels of manganese, crucial for plant growth and integral to multiple enzymatic processes, were significantly lower in glyphosate-tolerant corn and soybeans. "The indirect effect is through microorganisms in the soil. The bacteria are out of balance"—a situation that, he says, "does stimulate our disease-causing organisms. You wouldn't have a pesticide if it didn't do that. That's what makes it such a good weed-killer."

From the standpoint of soil, the genetic modification part of the glyphosate commercial program presents new risks and unknowns. There's been frighteningly little non-industry-funded research looking at what happens to us when the food we eat has been genetically modified. It's widely acknowledged that those who do conduct such research risk being run out of town. An example is Hungarian protein scientist Arpad Pusztai, who in 1998 publicly stated that genetically modified potatoes had caused slowed growth and lowered immunity in rats. Soon after he was suspended from the Rowett Research Institute in Aberdeen, Scotland. However, there's been even less—or at least less public discussion—about what GM crops do in the soil. One potential threat is "horizontal gene transfer," whereby spliced-in genes leap to other species. Once present in the environment, the fabricated DNA sequences can move around among soil organisms, forming new viruses and retroviruses, the consequences of which are unknown.

Mae-Wan Ho, a geneticist originally from Hong Kong and director and co-founder of the London-based Institute of Science in Society (ISIS), has raised a red flag about this phenomenon that rarely takes place in nature. In a 2011 report for ISIS she wrote: "Genetic modification and release of GMOs into the environment is nothing if not greatly facilitated horizontal gene transfer and recombination. It has created highways for gene trafficking in place of narrow by-ways and occasional footpaths that previously existed." She has called the introduction of transgenic organisms into the environment "much worse than nuclear weapons as a means of mass destruction—as genes can replicate indefinitely, spread and recombine."

Frightening words. Do we even know how worried to be? (I mean, there are just so darned many things to worry about.) Regarding such risks and permutations Huber says, "Our knowledge is so primitive. We're just stabbing in the dark."

On glyphosate, however, he's more clear: "We're starting to see a decline in productivity because of the effects of residual glyphosate in soil." Part of the problem with determining the chemical's persistence, he says, is that "the compound is so readily absorbed into the soil structure. Also, its degradation products may be just as toxic to some of the organisms." Then there's the disconnect between what happens in controlled laboratory conditions and in the field. "Most people don't recognize what glyphosate toxicity looks like," he says. "It looks like nutrient deficiency. Deficiencies of manganese, magnesium, copper, zinc—all those symptoms are also symptoms of glyphosate injury, because that's what glyphosate does."

I brought up the controversy over his letter to Secretary Vilsack. At the time, Huber explained, he chaired the American Phytopathological Society working group charged with managing a USDA plant disease recovery program. "I felt I had an obligation as a scientist and in my role" to alert the secretary "so we could do the research to maintain food safety and security. I intended it to be a very private letter." The attempts to damage his reputation didn't come as a surprise, he says. "When the letter was leaked and the problems that are common in the system were pointed out, it meant there was a lot of money and funding in jeopardy. A lot of people must have thought: we can't let this guy continue. That's not unusual when you look at the power structure we have. I would have been totally irresponsible if I'd neglected information readily available from other scientists—some of whom have been a lot more courageous, having lost jobs or been severely penalized. Science has never succeeded by burying the manifestations of unintended consequences. It has succeeded when we recognize them and deal with those issues. That's hard to do when you have a belief system that becomes a religion. In the minds of many, genetic engineering is a religion that saves the world, rather than something that threatens our sustainability."

I was able to speak with a scientist at Monsanto, David Carson, a twenty-five-year company veteran who works in environmental risk assessment. Monsanto's mission of sustainability, he said, is "all about yield. When you have enhanced yield you can feed more people. With more food per acre, less land needs to be committed to agriculture." He was not convinced by Kremer and Huber's findings on glyphosate. While there might be "subtle and transient effects on soil microbial life," he said, "within the next few weeks the herbicide will be degraded by soil microorganisms into natural components. Trait, soil, plant and herbicide interaction is very complex. We've done the field trials to separate those variables. The product's been around for more than 35 years. It's one of the most extensively studied and safest chemicals available as an herbicide."

Knowing what we know now about Roundup and Roundup Ready, or at least with a sense of what we should be trying to figure out, let's look at the global picture.

I'm not much of an intrepid third-world traveler, but my husband is South African so I've been to Africa a handful of times. On our last visit about five years ago we traveled to Mozambique. It was about the most enjoyable trip I've ever taken, mostly because the tone was set on our first stop, the Guludo Beach Lodge, an ecological, fair-trade resort. The lodge overlooks the northern edge of the country's magical Indian Ocean coast, graced by leggy palms and wooden dhows, the traditional sailing vessels that form airy triangles in the sea.

I remember two things from that trip germane to this discussion. While driving between destinations we often saw trucks hauling impressively vast hardwood trees along the dusty rural roads. Our driver told us these were Chinese traders: Timber buyers illegally pay locals to cut down the trees, which are shipped directly to Asia, threatening Mozambique's forests and providing no real benefit to the community.

Later, at our hotel in Ilha de Mozambique, a surreal, fortified enclave that was the earliest European settlement in East Africa, I saw a prosperous-looking Northern European man in earnest conversation with a series of various business types, loudly sharing photos of local properties from his laptop, from sparkling beaches and lush green slopes to once-grand stone houses on the island, making little attempt to keep his dealings private. Until this moment the world of commerce—or really the familiar world in any form—had seemed far away. When I mentioned this scene to the hotel proprietor he confirmed that real estate wheeling and dealing was beginning to stir, as were other business sectors. This was dominated, he said, by the Chinese. "The US and much of Europe sees its involvement with Africa as charity," he said. "The Chinese sees Africa as a market." And, I thought, recalling the trucks filled with logs, a source of natural assets.

With a growing world population, land, notably farmable land, is a finite resource. That visit to Mozambique gave me an inkling of where trends were going: Compared with the developed world, in Africa land and raw materials like virgin hardwood were a steal—and, so it seemed, there for the taking. And so we've entered an era of land grabs, an enormous shift of land wealth from common use or smallholdings to large corporate or governmental/NGO entities.

To offer a sense of scale: According to the International Land Coalition database landportal.info, since 2000 about 70 million hectares (173 million acres) of agricultural land in Africa—the equivalent of 5 percent of the continent—has been sold or leased to Western investors. Mostly large buyers: Of the 924 global land deals the organization has documented, 10 percent of investors account for 68 percent of all transactions.

The implications are huge. Sometimes land that has provided food crops to the local population is turned over to agrofuel plantations. One example is the introduction of jatropha in capital-starved countries like Mozambique, Ghana, Tanzania, and Namibia. In Mozambique it's been planted on nearly one-seventh of the country's arable land. The shrub, whose seeds produce an oil used for fuel, was said to grow well on marginal land and to need little water—beliefs that have since proven untrue.

The food crisis of 2008, with its rapid leap in grain prices and multiple food riots, sent many countries scrambling for backup sources. Cash-rich/crop-poor or high-population nations (China, India, and Japan in Asia; Bahrain, Kuwait, Saudi Arabia, and the United Arab Emirates in the Middle East) have been buying land for commodity crops (corn, wheat, soybeans, palm oil) as a hedge against food shortages and price hikes. Lester Brown of the Earth Policy Institute calls this "The New Geopolitics of Food Scarcity."

Who Owns the Soil?

I talked to Devlin Kuyek of GRAIN, an international nonprofit that in 2011 was recognized with a Right Livelihood Award (often referred to as the Alternative Nobel Prize). He said that many countries are finding it more economical to buy cheap land abroad—in a place like Mozambique land can be had for $1 a hectare—than depending on agriculture at home. "You can get more land in, say, Ethiopia, along with cheap water and export it back," he says. "It's the modern version of colonialism: it comes with investment agreements." A GRAIN report from June 2012 called "Squeezing Africa Dry" argues that the quest to secure water supplies is a subtext to agricultural land grabs in Africa, where already one in three people lack access to sufficient water supplies.

Then there's takeover of third-world farmland in the name of philanthropy. For example, aid and development organizations are seeding "biofortified" crops in Africa and Asia where people suffer from poverty and "hidden hunger" (nutrient deficiencies despite adequate supplies of food calorie-wise). The goal is to breed desirable nutrients such as zinc, vitamin A, and iron into staple crops like rice, cassava, and millet as a way of addressing persistent health problems like childhood blindness and anemia. These programs are being deployed rapidly and on a vast scale, one recent effort being the introduction of iron-rich beans in Rwanda. Biofortification receives significant funding from the

Rockefeller Foundation and Gates Foundation, the latter of which has financial and personal connections with Monsanto and Cargill, another agricultural giant. Donors to the biofortification nonprofit Harvest Plus include USAID, the World Bank, Syngenta Foundation (associated with the Swiss chemical firm), and the International Fertilizer Group.

The introduction of biofortified crops has not gone smoothly. Syngenta's Golden Rice, genetically engineered to produce beta-carotene, a precursor to vitamin A, is now in field trials after several years of delay because of multiple patent disputes. A BBC report noted that malnourished people may not absorb the beta-carotene from the rice without a balanced diet that includes the type of traditional foods that commodity crops like hybrid or GM rice put in jeopardy. It's also been found that to get the benefit from fortified rice, young children would have to consume six pounds of it a day. When it comes to malnutrition, addressing wealth disparities and improving access to a wide range of foods seems a more direct way than fiddling with genes and traits. As Hans Herren of the Millennium Institute told the BBC, "We already know today that most of the problems that are to be addressed via Golden Rice and other GMOs can be resolved now with existing and tested means, with the right political will."

There's also the thorny problem that people may prefer traditional produce to, say, bright orange sweet potatoes or rice that has an unfamiliar texture. Thus, a large proportion of effort and funding goes into marketing these products and educating "target populations" as to their merits. Citizen groups and regional organizations such as the African Centre for Biosafety have sounded the alarm that biotech "solutions" to hunger threaten food sovereignty and leave people vulnerable to the pricing and products of multinational corporations. Populations resist—Zambia and Angola have outright rejected GM food aid—but there's great pressure (the word *bullied* is often used) to accept commercial biotech. Biofortified crops and the many other agricultural development projects in the global south, such as the US government's Feed the Future initiative, are also inevitably based on mono-cropping and biochemical inputs.

Agricultural land grabs can be seen, to borrow author Naomi Klein's apt phrase, as a kind of "disaster capitalism," exploiting calamities like droughts and hunger to intensify biotechnology in the third world and therefore perpetuate economic dependence. Local residents are promised jobs, food security, and community investments such as schools and hospitals. However, the more likely results are people displaced from their land and livelihoods, communities broken, increased vulnerability to food shortages, and conditions ripe for political unrest. These forays are couched in terms of "integrated value chains," "priority commodities," and "catalytic philanthropy." But ultimately the quality of food produced depends on the integrity of the soil, something not noted in tallies of yields and profits or in lofty developmental missions.

Land grabs are happening all over the world, including Eastern Europe and Australia, says Hans Herren. "People with money realize that people will need to eat no matter what. Only three percent of the world's land is arable land. With growing demand for food, fiber and feed, the value will go up. The idea is to invest now, even in land that isn't currently productive. Corrupt politicians who may grant long-term leases to equally corrupt banks or corporations are responsible for many of these deals that in the end affect the poor. They come in and kick the people off the land because there is no place on earth where there isn't somebody at least part of the time, such as grasslands where pastoralists graze with their animals once or twice a year. The corporations behind the deal eventually bring in their own people, machines, fertilizers and seeds. Soil fertility will go down as under this type of industrial production they're not using organic agricultural principles, or crop rotation practices with legume crops that regenerate the soil. When you invest money in such land grab schemes, you want to break even in five or so years." There are exceptions, he says, but usually "there's no intention of making sure that this is for the long haul."

Herren, who has won several awards for work in ecological science that promotes living standards, particularly in Africa, says he opposes GMOs and biofortification as ways to bolster nutrition. "If you look at crops from before the green revolution, they were nutritious," he says. "Breeding has raised the starch and water content. With high-yielding varieties we have increased crop yield but lowered the nutrition."

He expresses concern about narrowing the genetic base, as farmers are "forced to give up diversity for a few varieties and simplified systems. This increases vulnerability to insects, diseases, climate variations such as too much water or not enough water. We have the [GMO] technology, so it's as if the technology is looking for a problem to solve. With it, as earlier with pesticides and herbicides, we treat the symptoms rather than tackling the underlying causes. For example: what causes nutrient deficiencies, and pest, disease or weed outbreaks? Following this path, we eventually end up on a treadmill of pesticides, fungicides, herbicides, synthetic fertilizers, GMO seeds without solving the problems permanently."

These words apply to Iowa cornfields as well as the Limpopo Valley in southern Africa, to the cultivation of fancy mesclun for urban bistros as well as starchy cassava in African villages. Chemical inputs and biotechnologies can mask poor soil quality for a period of time. But they also set up a treacherous course of imbalance and dependence.

Chapter Eight

Floods, Drought, and the Grasslands, LLC, Experiment

Floods do not begin as floods. They begin with drops of rain hitting dry earth.

—Allan Savory

"Rain doesn't only fall from the sky," I suggested. "It also falls up from below."

—Masanobu Fukuoka, from *Sowing Seeds in the Desert*

EVEN IN AN ERA OF SATELLITE NAVIGATION and Google Maps, the Cinch Buckle Ranch, which straddles Powder River and Carter Counties in southeastern Montana, is tricky to find. In a surveyor's drawing the dirt roads get thinner and squigglier and eventually run to dots. That off-the-map sensation truly hits home when I'm driving around looking for it close to 5 am and the sun is lifting in dusty orange bands and I'm wondering if I'd missed the old windmill or the clump of corrals that would mark the left turn and, several miles after I thought I should have reached it I'm beginning to suspect I'm lost.

Still, it's a beautiful drive, with the slowly brightening late-May sky and sightings of pronghorn antelope, fleeter and huskier than their more common white-tailed cousins, zipping across a hillside or in a frozen pose. I finally see a rise tall enough for cell reception and climb out to confirm that, yes, the turnoff was fifteen minutes behind me. I'd left Broadus, a ranching and hunting outpost that styles itself the "Wavingest Town in the West," at four thirty because Cinch Buckle manager Ron Goddard said he'd be moving some cows from a spot miles away from headquarters and if I wanted to watch this I had to catch the early show. The ranch is run according to Holistic Management, so the grazing schedule is essential. And here not only am I off course and bothering my host at odd hours, but I'm also holding up the cattle.

When I arrive at the base, Ron and his wife, Kathleen, are suited up and on their horses. Ron asks if I ride. He catches my stunned expression—I'd only ever done group trail rides on super-docile, usually elderly horses—hands me a set of keys, and says, "Here, you take the pickup." As there's no time to lose I jump in and follow the couple on horseback, down into a dry streambed and through an open fence and out into beautiful open pasture with a backdrop of hills pushing back toward higher, sharper hills. We go on a good distance, into and out of culverts, through high grasses. Even driving a vehicle feels athletic; I could only imagine what the riding must be like. The two look so regal on horseback, straight-backed and kitted up in fine leather gear—tooled boots and fringed chaps—moving in and out of silhouette as they weave across fields. I feel nostalgic for cowgirl dreams I never had and westerns I'd heard of but never seen.

As we pass over one broad ridge the cattle, mostly black and in pairs or groups of three, come into view and the horses pick up speed. We're now in rodeo mode as Ron unfurls a whip and swishes it this way and that, sketching sharp lines in the air. Kathleen charges in from a side angle, her horse kicking fast and dustily and high. The cows step more quickly, mooing irritably to show they're not happy about being told what to do. I'm farther back among the stragglers and realize that without thinking I'm "herding" cattle with the truck, nudging them to move faster and toward the gate. It feels like a kind of instinct.

When most of the animals are through Ron sidles up to me and motions toward a hill. This is where I should wait with the truck while he and Kathleen go to work. Their task now is mostly "cleaning up," he explains. The gate between pastures has been left open so that calves can find their mothers. We're toward the end of the calving season, and "when the cows calve it makes Holistic Planned Grazing more difficult," he says. "You get about an hour right at gray dawn" to move them as a group. Apparently the calves nurse and then lose track of their mothers if you're any later than that. Which we were—thanks, in part, to me. So Ron and Kathleen prod and chase lost calves—and cows moseying about looking for errant offspring—until they're all past the gate and reunited in the next pasture. It is also, I realize, a somewhat noisy operation as cows complain loudly and continuously when they can't locate their calves. I take a rest in the truck, taking in the sharp, dry scent of the sage that covers the hill and listening to the sounds of birds and the wind whipping against the partly open window. For a long while there is no cow, or man, in sight.

According to Ron, there are seven hundred cow–calf pairs in the section they were working, three thousand on the property. "We're in a pretty severe drought. If things don't change we'll have some challenges," he says. The year before, 2011, they had flooding. "It washed out fences and caused some severe erosion in places, but overall it was beneficial. Average yearly precipitation here is fourteen inches. We had eleven or twelve last April and May. If we could have had half of that last spring and half of that this spring, that would be my vote. But we don't get to choose."

I'd come to the Cinch Buckle Ranch to observe Holistic Management, the use of livestock to restore land function, in action. I'd read and heard so much about the model; it was my reporting on Holistic Management that drew my attention to soil as a hub for the ecological cycles whose disruption is leading to environmental crises—and whose restoration can bring aspects of our planet back into balance. In the course of writing these chapters, I've seen that there are many routes to repairing the carbon, water, energy, and nutrient cycles; different ways to put spokes back on the wheel. Whether we focus on biodiversity (as does Gene Goven in North Dakota) or carbon (like Christine Jones) or water (per Michal Kravčík), the other components will follow. You could say, then, depending on your theoretical or practical orientation, that biodiversity (or carbon or water or cover crops) is driving the processes.

With Holistic Management, it's livestock that get it all going. Through their behavior—grazing, trampling, leaving waste—they set multiple biological dynamics in motion. I like cattle. I like their absurd, stolid, clueless bovine-ness. I like the disarming, seemingly preposterous notion that while we futilely chase technological and policy rainbows, a key to our problems could be something as prosaic or random as the browsing of a cow. So I wanted the chance to witness the cattle working their miracles, and to meet the people who make Holistic

Management happen, the ones who do the work of moving cattle and monitoring changes in service of improving the land.

Launching a "Brown Revolution"

The Cinch Buckle Ranch is part of an ecological/business experiment called Grasslands, LLC, a joint venture between the Savory Institute and investors John Fullerton (whose firm is Level 3 Capital) and the family office of Larry, Tony, and Michael Lunt (Armonia LLC). Fullerton, whom I've had the chance to meet a couple of times, is a former managing director of J. P. Morgan. After leaving the firm in 2001, Fullerton began grappling with questions of economics and sustainability. He has since formed the Capital Institute, a nonprofit forum on the role of finance in a shift to a more "just, resilient and sustainable" system. In 2008 he embarked on a correspondence with Allan Savory. After much discussion and sharing ideas about the need for a "Brown Revolution"—an agricultural transformation centered on restoring soil health—they sketched out a plan for a custom-grazing business grounded in Holistic Management.

In 2010 Grasslands, LLC, was launched as a "triple bottom line" enterprise with the conjoint goals of creating a high-quality product (well-nourished beef cattle), generating equity and financial return to investors, revitalizing rural economies, sequestering carbon, and regenerating land on a large scale. Here the goals of improving land and making a profit would not be mutually exclusive: Holistic Planned Grazing requires a lot of animals, and in turn bolsters the carrying capacity of the land, sometimes two to four times. The more animal impact, the better the land—higher soil carbon levels, greater biodiversity, better water infiltration—and the more animals it can feed. This means greater income and a boost to local economies. It's "impact" investing on many levels.

The company first purchased two ranches (the BR and Horse Creek Ranches, fourteen thousand acres combined) near Newell, South Dakota, land then under absentee ownership that was producing below its potential. (Jim Howell, Grasslands, LLC, co-founder/CEO and a longtime practitioner of Holistic Management, says the West is full of "trophy ranches," tracts of rangeland with animals scattered about, bought primarily for the romance.) The Cinch Buckle Ranch's thirty-nine thousand acres, much of it public Bureau of Land Management and state grazing lands, were added in early 2011, followed by another large (fifty-three thousand acres) eastern Montana property in 2012.

I decide to focus on the water cycle. Aside from higher profitability—the result of increased stocking rates—the most visible and dramatic improvements achieved via Holistic Management often involve water. (I have an image from the Africa Centre for Holistic Management of elephants happily watering on the newly restored Dimbangombe River, as opposed to having to travel to pools.) Also, every ranch treads the fine line between flood and drought, a surfeit of water all at once or a thirsty lack of it, with the specter of opportunistic grass fires lurking in the background. In "brittle" or seasonably dry areas, like the American West, the more frequent worry is drought. A livestock operation rises or falls on the availability and management of water, the vagaries of which are so unpredictable that few private insurers cover droughts, floods, or fire.

After a bowl of chili at the Cashway Cafe in Broadus the night before my early-morning drive, I'd walked up Highway 212 and noticed the sign at Copps Hardware, which read: PRAY FOR RAIN; WE'VE GOT TWINE. And this was well before the extreme drought and the rash of uncontrolled fires that plagued the area—and much of the West and Midwest—in subsequent summer months.

Ron Goddard, who's been in Holistic Management circles since the mid-1980s, is originally from Kansas. He came to Montana when he was just out of high school. "My family raised cows and grain," says Goddard, who's a fairly big guy and sports a brushy, graying mustache and broad straw hat. "I didn't like farming. Still don't. When I was old enough to get out from Dad's sway I went around the country." By the time he met Allan Savory he'd set up a custom branding business in Montana, working out of a horse-drawn wagon: "I kept hearing about planned grazing and went to one of the early workshops." Which inspired him to make the switch to land and livestock management, freelance-style. While he's tried to follow Holistic Management principles, "it's been a struggle. I've been leasing land or working for people who hadn't believed in it. Sometimes I did it under the radar."

He and Kathleen have moved from ranch to ranch, living in eleven different states, and raised three children, schooling them at home. Jake, the youngest at eighteen, works with them at Cinch Buckle. "This is his bread and butter—horses and cows," says Goddard. The older son shoes horses and their daughter, recently married, works on a ranch with her husband.

Goddard gives me the lowdown on the particular plot of land in front of us. "The previous owner had it five or six years and ran fifty to sixty head of cows year-round in this pasture," he says. "The creek bottom is overgrazed, and noxious weeds are taking over. In this kind of environment most land is undergrazed, so it's hard to change succession."

By "succession" Goddard means the evolving mix of plant species in a given area. In well-functioning land, the plant community moves in the direction of complexity: a mix of varieties in which plants fill different ecological niches. This diversity creates resilience, the ability to withstand difficult conditions like an onslaught of pests or a drought. Land can become stalled at a low level of succession, whereby only a few kinds of plants dominate. Such land is inherently unstable and vulnerable to external threats. Succession in the animal population parallels that of plants, so that the biodiversity of plant life will be echoed in the variety of animal species. Animal impact promotes succession in several ways: Grazing exposes the growth points of different plants to sunlight; trampling helps decompose litter, breaks hardened soil, and pounds seeds into the ground so that dormant species can be established.

After a single year it's hard to see much land improvement, but Goddard says they're doing a better job of utilizing resources. He wants to keep a lot of cows, now in three herds belonging to three different owners, but expresses concern about scant spring rains. "We may need to de-stock if it gets severe enough," he says. "What we want to do over time is improve the soil so that less rain does more good—that it's more effective." He notes that the fields near the house that had been farmed "have lots of bare ground. A drop of rain does more good in natural rangeland that's undisturbed except by livestock. Hopefully by adding rest periods and shorter grazing periods, more plant [varieties] will come in. We'd like to see more forbs and perennial grasses. With the level of rainfall we have, there will probably always be lots of western wheatgrass"—a variety that is fairly drought-tolerant.

The area was not always so dry, Goddard learned. During his first spring at Cinch Buckle he lived on his own while the family stayed at their home near Billings. "I went to the local library and read self-published memoirs," he recalls. "There were a lot from the 1880s to 1930s and a comment I read several times was, 'Back when the cricks used to run.' Now they just run in the spring, maybe a little longer if it rains a lot."

Goddard says that from the late 1800s on this region—the corridor that runs from eastern Montana down through Belle Fourche in South Dakota—"was like the highway for Texas cattle, the stop between Texas and Miles City, which used to be Fort Keogh, and is now a little Las Vegas." I'd passed through Miles City the day before, and Route 59 was strewn with casinos, strip-mall-style, with western gold rush names like Golden Spur, Silver Star, Lucky Lil's, and Gold Dust. It was a big homesteading area, he says, "initially big cattle outfits. It was always big ranching country, cattle and sheep." Immigrants from England and Scotland in particular brought cattle to the land. "The grass was free, and the country wide open," says Goddard. "That didn't last long."

Once we're back at the house—at an hour I'd usually just be getting out of bed—it's chore time. Kathleen asks: "Would you like to feed some calves?" This sounds like something I can handle so I say, "Sure," and follow her into the calving barn, near the stable and paddock area where they keep their thirty horses. She hands me a repurposed plastic bottle full of milk and directs me to the calf I will feed, a near-black three-month-old baby cow with limpid eyes and out-flung ears. She's a good feeder, determined to find more milk after she's drained the bottle. There are three calves in the stalls, and I ask why they're not out with the others. Kathleen explains that these were orphaned calves.

"During the Texas drought last summer, some ranchers sent their cattle up here to graze rather than buy expensive hay," she tells me. "In dry weather, there's a shortage of grass. Some mothers were thin to begin with and a number of cows died. They'd been through almost a year of drought and then came here in the fall. In the winter they started having calves. Though we had a mild winter there was still a week that didn't get above zero and many couldn't make it. We lost a lot of calves and quite a few cows." She's been feeding the ones without mothers, a group now down to three. "They'll go out with the herd once they're old enough to eat grass," she says.

Ron, who's come into the barn to return some tools, overhears us. "Everything that could happen to those Texas cows did," he says, shaking his head. "There was a one-month-old whose mother drowned in a creek after getting stuck in the mud. I write a report once a month to send to the owners. I've had to say, 'The good news is you have lower numbers to feed. The bad news is that it's because animals died.'"

We head back to the house and eat breakfast, herb omelets from fresh farm eggs, while sitting on tall chairs around the counter. "In the place I grew up, and virtually every place I've been, the one thing I hear is: 'When Granddad was here grass grew up to your stirrups and the crick used to run,'" Ron says. "We like to blame it on climate change or too many animals. But that doesn't hold water." In his younger days, he says, he "just wanted to ride horses. I wasn't thinking of why." It wasn't until hearing Savory's ideas that he understood the extent to which fluctuations of water are a reflection of how we treat the land: whether it's managed toward resiliency (covered ground, high levels of organic matter, good soil structure, all of which retain water) or degradation (bare ground, depleted soil, crusted surfaces, conditions that lead to evaporation and runoff). The idea that running livestock—specifically, running livestock the way buffalo or wildebeest move across the savanna or plains—could invigorate land was a revelation, and has since inspired the couple's lifework.

"This short-grass prairie is pretty high on the brittleness scale," says Ron. "It's similar to where I grew up in western Kansas, which is right on the one hundredth meridian." That land, he says, has been "farmed up." From childhood he recalls "a prolonged drought in the 1950s. I remember dust storms when my mother wouldn't let us go outside. She was afraid she couldn't see us in the dust. I went to a neighbor's place after their house had blown away in a tornado. They were safe in the basement. Their telephone was found six miles away—in those days, they still had the number on the phone. Now that country's dried up." Both in Kansas and Montana, he says, "it seems the small towns continue to shrink. There was [once] a school right down the crick, about a mile from our house, the Moore School. You see it on old topos. Nothing left there now but a trace of a building." A bit farther away, Powderville, which I'd unknowingly driven through that morning, was once a thriving town; its population is now listed as seventeen.

With dry terrain, wildfire is always a threat. "There are some hellacious lightning storms that come through here," Kathleen says. Ron adds, "Last year [2011] it was wet and we had a huge amount of growth. Then it got real dry. Fire's always in the back of your mind. When you see smoke when the conditions are right, you don't worry whose land it is, you go to get help. There were 120 different fires last year in southeast Montana. One thing it does is take out all the good [plant] cover. It makes land like this more susceptible to erosion."

The loss of plant cover after fire also alters land's ability to hold carbon and water, and support soil microbial life. Peter Donovan has written that fire speeds up the carbon cycle, adding that "overall biodiversity is compromised, as carbon is recycled through combustion rather than decay which is much slower and requires and feeds an enormous range of organisms."

The pressure to make the land profitable is a constant backdrop. And the Goddards are continually aware of how much money is involved in the operation, in part due to the skyrocketing price of ranchland. This a reflection of economic uncertainty, says Ron. "A lot of people don't want their money in the stock market. They want something real." (A few days earlier when I was in North Dakota, land inflation was a big topic of concern, too. At Black Leg Ranch near Bismarck, owner Jerry Doan told me, "Granddad bought this land at $1 an acre. Other people wouldn't risk that money. Today it's up to $1,900 an acre.") It doesn't bother Ron that neither the land nor the cattle is "his." "I don't care whether we own it," he says. "If we can manage the land well, that's our goal."

The long time horizon often necessary before seeing improvement under Holistic Management is an ongoing test of commitment. "We were in one place for ten years and saw positive changes in the percentage of covered ground and the number of species, but it was years before we started seeing anything measurable," says Ron. He and Kathleen accept, and even expect, this: "It's a slow process in our mind." Areas that had eroded in ten-foot-deep cutbacks were starting to stabilize, he says. "Then we lost the lease. The cattle market got so good the owner wanted back in."

That ranch, in central Montana's Clarks Fork Valley, was easier terrain to work than Cinch Buckle. "This is a hard environment," Goddard says. "It didn't rain June of last year till April. In degraded land, soil becomes less of a sponge. It runs off, eventually into the ocean. There are springs marked on maps from the '20s and when you ride to that place now, it's just a bit of mud." Another challenge is the type of soil that dominates much of the land: "gumbo," a heavy, sticky mineral soil that can get kind of gummy when wet. In *Backpacker* magazine, writer Tom Shealey described its texture as "somewhere between quicksand and Silly Putty."

When Goddard first learned about Holistic Management and its framework for the interplay of livestock, land, and water, his response was, "This makes so much sense—everybody's going to be doing it." He found, however, that in most ranching circles it was a tough sell. "Allan Savory said that first your neighbors will think you're crazy. Then when you're doing well they'll drive past your place and not look," says Goddard. "We don't hang out in the bars and coffee shops in town around here so that people can tell me how stupid we are. We help our neighbors and they help us and we're social and go to church. But we don't tell anyone they should do this."

Kathleen concedes that migrating from ranch to ranch can be disruptive and lonely; she misses her friends in central Montana, bonds built over many years. I see that there's a toughness to Kathleen, who's petite and athletic and has thick brown ponytail-able hair, which I imagine sees her through the difficulties of this life. Rather than complain, she finds a bright side: "I've been blessed in that I've never had to go to town to work."

For Kathleen, it's all about the horses; she can't imagine a life that isn't centered on riding. If Ron landed in Montana upon fleeing the farm in Kansas, Kathleen came from Louisiana to immerse herself in equine country. "I came with my sister at age eighteen and we both ended up marrying ranchers," she says.

She tells me about their adventure in herding horses. A friend, Warren Johnson, had bought a thousand Premarin geldings. These are horses whose mothers' urine had been harvested to make estrogen replacement therapy; the female foals are raised for their role in Premarin production as mares, which leaves the geldings. He transported them from Canada to Roscoe, Montana, to prevent their sale to meatpackers. The rescue process involved much paperwork and a team of thirty riders to lead them down the highway to their new home at the Lazy El Ranch, which Ron then managed. "We ran eight hundred horses for two years to holisticate with them." She smiles, acknowledging her conscious use of *holistic* as a verb. "They were pig-fat, ready for slaughter. The geldings, the boys, are seen as a by-product [of Premarin production] and are shipped to feedlots." The market for horse meat is abroad, in countries like Japan, France, and Belgium where it is considered a delicacy.

Premarin mares are reportedly confined in pens for months at a time and deprived of water, which prompts greater estrogen production. The stressful conditions the geldings had endured in the feedlot where they'd spent most of their lives were still evident. "Their tails were bitten off, they were so bored," Kathleen recalls. "They would herd up. They didn't go slow and we had to train them to stop when they came to a wire fence. Initially they would pile up against it until the leaders were pushed through. You see, they had never been free. They'd spent their lives in small, steel pens. Eventually they wouldn't be spooked by a butterfly or bunny rabbit—things they'd never seen before." The horses have since been sold to farms and ranches and outfitters, as workers, Kathleen says. "There was a lot of draft blood in many of them, which made them desirable for outfitters to use as pack and saddle horses in the mountains."

Making Use of Rain

After Cinch Buckle, I leave for Belle Fourche, South Dakota, a former trading outpost that still holds regular livestock auctions and where I spend two nights. I take a sightseeing day in the nearby Black Hills, and it starts to rain. It's surprisingly chilly when I get out at this or that switchback along the Needles Highway to check the view. Mount Rushmore is locked in a wet fog, its famous stone faces invisible. I give it a pass.

The next day I head over to Horse Creek Ranch, an eight-thousand-acre tract purchased by Grasslands, LLC, in 2010, about forty-five minutes east of town. As I drive we get another dose of rain; I go through the town of Newell, once known as the "sheep capital of the world," which still features the annual Ram Sale, where I see a man hosing down his red truck to clean up after the shower. I spot a band of blue in the horizon. The sky clears. It's like a skywide film being peeled back bit by bit.

I arrive at the appointed gate in good time. Between the civilized hour—late morning—and the clearly marked grid of straight, broad dirt roads, the ranch is easy to find. A few minutes later a black truck rumbles up and out steps a very tall, blond man in a white western hat, a bright-eyed woman with copper-colored hair and straight bangs holding a baby at a casual angle, and a round-faced boy of about five darting toward a tree, now crouching near the truck, moving about to elude the indignity of adult scrutiny. This is the Dalton family: Brandon, who manages Grasslands, LLCs, two South Dakota properties, Brandy, Emmett, and Garrett, who is not quite one.

They make room for me in the truck, and I comment on the rain. The pastures are flushed green; I'm glad for my good timing. Brandy shrugs. "It's about a quarter inch," she says. "It's not a game changer." I feel disappointed, as if it were my own offering that had been rebuffed.

Brandon, twisting the steering wheel hard to minimize the bounce over uneven turf, says, "It keeps the grass going so it keeps it green. So that when the one-inch rain comes it can do something. If not, the land couldn't do anything with it." If it fell on dry land, much of the rain, he says, would simply run off.

When people consider the growing potential of a particular environment, they often look to the amount of rain it receives. Holistic Management emphasizes making effective use of whatever precipitation comes down. To Brandon, this is an inherently optimistic model: We can't force rain down from the sky but we *can* take measures to ensure that the rain we get is used well. He embraces this philosophy. "The thing about Holistic Management is how uplifting it is," he says. "When you look at things holistically you can find a lot of hope."

As he describes it, the model is an ongoing exercise in hope and action. "Holistic Management is about making a plan, monitoring the plan, controlling deviations from the plan, and then replanning," he says. "When you make a plan, you assume that something will need adjusting. This takes the pressure off from trying to be always right. I don't worry about being wrong. I want to make sure I have a plan so I can adjust." The rain pattern you get in a season is impossible to foresee, so almost by definition there will be tweaks along the way.

Both Brandon and Brandy grew up in Wyoming, Brandon in the western part of the state and Brandy in the east in Hulett, near Devil's Tower Monument. Brandon is a direct descendant of the Old West train-robbing outlaw family made famous as the Dalton Gang; Emmett is named for the brother who lived to tell the stories. The two met at the state university in Laramie. Brandy studied agricultural education and nursing; she's a licensed EMT, which, she says, "is good when you're living in remote places." Brandon earned a BS in biology and later a master's in zoology at Washington State; he has a bit of the scholar about him, a thoughtful stillness as he considers a question. Brandy is the outgoing one. She radiates buoyant energy, a hint of fun.

Brandon was introduced to Holistic Management through Brandy's family. "My dad [Nick Bohl] had taken on management of thirty thousand acres of land," she says. "His goal was to be the best manager he could. He took a seminar with [Holistic Management educator] Roland

Kroos. Kroos helped him implement a plan and made that ranch more profitable than it had ever been." Brandon says, "The increase in wildlife from Holistic Management piqued my interest as a biologist." As he began practicing, "I realized I had a passion for grazing—the habitat you can create with it."

Brandon stops the truck and we get out near a dip in the landscape. The view, mostly green with strips of yellow, is vast, stretching in soft undulations toward the horizon. And this is one of Grasslands' smaller ranches.

"I picked this spot where you can cross the creek," he says. Apparently that minor indentation in the land is a creek. "Last year at this time it was full of water. When water comes down too fast and the bank is not stable, it's cutting through the soil." As he points and explains, I start to see the place through his eyes, as a kind of legible diagram or document. I follow the diagonal line that rushing water carved in the creek bed. "I'd like to see some bank stabilization and fill in these cuts. Above the cuts, the water dispersed. That kept more on the ranch and meant that flooding downstream was minimized." He mentions the 2011 floods along the Missouri River, which caused tremendous damage throughout the Great Plains, and says, "Those floods were coming from places like this."

Brandon crouches down and brushes his hand over the grass. "We want to encourage grass like this—cordgrass, a plant with deep roots," he says. "Like willows. This ranch has one patch of willows. I'd like to see more healthy riparian vegetation on this creek, willows and perennial grasses with all those microscopic roots as opposed to our weedy species, which are taprooted. Those are better than nothing, but you need the fibrous roots to stabilize soil." He muses about how to make this happen, perhaps bringing in thousands of willow cuttings to plant, then pauses and sighs. "Here at South Dakota Grasslands, we've got myself and part-time help. The time scale will depend on the resources. Grazing will be pivotal, though. Willows will come in with well-timed grazing and recovery periods."

I stoop down next to Brandon and see an all-too-familiar plant, one that's irritated my skin more times than I could count: Canada thistle.

He plucks a prickly leaf, which he folds up ragged-edge-inward and eats. He encourages me to do the same. It's not as bad as you might think. It has a kind of generic-green taste. "Cattle do tend to eat it," he says. "If they're spread out season-long, they won't. Most of this is western wheatgrass, a cool-season perennial grass. The landscape here is not very diverse. Livestock will look for something different. One species rarely provides all of what an animal needs."

He stands his full six-foot-plus height and indicates the sweep of this pasture. "No one's grazed this since August last year," he says. "Last year, in late May, the grass was shin-high. A wonderful thing about this region is even in a dry season you reliably get spring green-up. I grew up at seven thousand feet, cold desert. This region is less susceptible to real droughts. But I know it gets real dry. The locals say, 'Just you wait.'"

Brandon's family is in southwest Wyoming, an area now under rapid natural gas development. His father and mother are, respectively, a social worker and middle school counselor (though his father just retired). "I'm still a biologist," he says. "I probably do more biology now than most biologists." In addition to managing the two ranches, he's also working to set up research projects.

We get back in the truck and Brandon says we'll head back for lunch. Unlike the Cinch Buckle, at the Horse Creek and BR Ranches they do their herding by four-wheeler. Brandon explains that it's more efficient, as they don't have spots that can only be reached on a horse. He says this somewhat shamefacedly, as the horseback version ranks so much higher on the authentic cowboy scale; few think *Wild West* and picture an all-terrain vehicle.

"Look at the fence line," he says. "Ours is green and across it's brown. What we've been able to do is put litter on the ground. At the neighbors' it's still standing." They accomplished this, he says, "mostly running yearling cattle. On the two ranches combined, in 2010 we ran twenty-three hundred cattle and in 2011 we had thirty-six hundred. These bigger herd sizes tend to knock down litter pretty well. If we get an inch of rain, the land will really hold water. Plants stay in contact with the soil surface. That allows the microbial life to break it down. The green/gold you see is from last year's growth. The gray, mixed in, is from the year before that. There's not so much old growth that it's blocking sunlight, but enough to hold the snowmelt."

Home turns out to be a camper in a field. Waiting for us is Emily Jerde, who helps part-time. At twenty, she's studying violin long-distance via the Berklee School of Music and training to be a midwife. Her family runs buffalo using Holistic Planned Grazing, so Emily is well acquainted with the routine (if, indeed, it ever is routine). Brandon stresses how lucky he is to have her there. "As for people excited about Holistic Management, there's no shortage," he says. "People with the skills are another story."

It's a tight squeeze, but we gather around the camper dinette and Brandy dips into the Crock-Pot and serves the most amazing stew. It's elk stew—from an elk Brandon harvested back in Wyoming the previous fall. "Elk is said to be sweet meat," she says, adding that this one has carried them through the last several months. I empathize with what I imagine is the challenge of keeping a growing family equipped on a remote ranch with minimal space, but she seems unfazed. She puts Garrett down so he can speed-crawl about the cabin.

As happened at the Goddards, we lament the exodus of young people from the plains; it seems a theme we can't escape. "What we need to do is instead of the best and the brightest leaving, get them to come back to the ranch," Brandon says. Brandy adds that young people may be tempted to leave since they're attractive to employers: "In a job interview as soon as they find out you were raised on a ranch, you're hired. They know you have a lot of skills and can work hard."

The other perennial topic is the threat of wildfire. "It's a pretty high risk if you have a lot of fuel built up, standing stuff five years old" that hasn't been reincorporated into the soil, either by cattle (trampling) or microorganisms (decay), says Brandon. "It's always windy. In 2010 a thunderstorm passed by overhead and two weeks later we found a small burn patch, about half an acre. It had probably rained itself out, but that could have burned up a lot of acres very quickly."

"If we're successful with what we want to do we'll increase fire resistance. A lot of fresh growth means moisture," says Brandy. "You can also use grazing to create firebreaks. And build the southeast summer wind pattern into the grazing plan. You can mitigate some fire danger that way. If we can improve our water cycle so that we're retaining more water in the ground and in the plants—that's what we're striving for."

This management approach, she says, can also be applied to woodlands—where the largest and most intense fires occur. "A lot of forest fires are fueled by dried old grass or an overly high density of trees," she says. "Grazing can keep vegetation vibrant and green in the understory, to reduce fuel loads and improve the water cycle."

This is the family's second season in the camper; they haven't been overwintering cattle at the South Dakota ranch. "This place will never be good winter country," says Brandon. "We're facing northwest, where the winds are. In intense weather events, cattle can die." Come November 1, the Daltons will be at the new Grasslands, LLC, operation, Antelope Springs Ranch, a fifty-three-thousand-acre property in east-central Montana north of Miles City. "It used to be called No Creek Ranch, but we want to name it after something it has," Brandon says. He's excited about the move, he says. "It's a big transition, a good permanent place with seventy-five-plus square miles to get to know." I pause, trying to wrap my head around the notion of seventy-five square miles. "There's 640 acres in a mile," he says. "That's big. It might be closer to eighty. I love exploring new country, big country, and there's a lot of it out there."

Brandon reflects on his South Dakota time so far. "We grow more grass at this point in the season than where I grew up in an entire year. It's working out well this year, but last year it was seventy-five hours a week with no days off for months on end. The thing about cattle is that it scales up nicely. Grazing management for five hundred cows [is] much the same amount of work as fifteen hundred cows."

Brandy listens to the two of us talk about scale and strategy, grazing and planting, and puts in her two cents: "With Holistic Management it almost doesn't matter what you're doing as long as you're paying attention."

For Want of Water

Allan Savory has said that floods and droughts are man-made. At first glance this seems a bold, even brazen statement. But here's what he means: When it comes to the water cycle functioning in a landscape, the condition of that land is as important as what descends—or not—from the sky.

In an essay on the summer 2012 drought, Savory wrote:

> In areas covering most western U.S. states suffering from drought you can, as I have done repeatedly, stop anywhere and sample the land from the best conventionally managed ranches to wilderness areas. Commonly you will find that no matter how good the grassland might superficially appear that anywhere from 50% to over 90% of the soil is bare between grass plants. This state of affairs guarantees ever-increasing frequency and severity of droughts just as is so tragically being experienced. I have found almost all range scientists, wildlife scientists, foresters and ranchers as well as conservation organizations believe this condition is natural. Nothing could be further from the truth. It is entirely unnatural and man-made.

The reason this happens, he says, is that there are too few large herbivores (mainly livestock) on the land—and those that do exist are showing unnatural behavior given the absence of pack-hunting predators. This leads to over-resting the land while overgrazing plants. Plus the overuse and misuse of the tool of fire.

Savory's is an unconventional perspective, but ultimately a very empowering one. His analysis is that the things we humans have collectively done to the land have interfered with the effectiveness of the water cycle. These activities—farming, allowing for overgrazing and undergrazing, clearing trees and other vegetation, over-relying on fire as a management tool—have left land less able to absorb water. (Bringing in the New Water Paradigm model, I'll add building on or paving over land, which also seals soil and impedes water infiltration.) This leads to drought and runoff, which then causes erosion and floods. However, this situation can be reversed by managing land in such a way that the soil's ability to retain water is restored.

It's a stance we might want to reflect upon, for floods, droughts, and wildfires (a greater threat when conditions are dry) affect more than just those who work the land. I've mentioned Hurricane Irene that swept the East Coast, killing sixty-seven people and causing more than $15 billion in damages. In Vermont, this deluge followed spring floods bad enough to warrant a presidential declaration of disaster. That same summer, Thailand saw severe flooding during the monsoon season, disrupting the lives and livelihoods of people in much of the country, including in Bangkok. We'd hosted a Thai exchange student, Naim, a few years back, so I'm somewhat tuned in to news about Thailand. Due to high waters, the school Naim enrolled in, near Bangkok, was suspended midterm and its resumption was delayed three times. It's been called the worst flood in the world to date—Noah notwithstanding—in terms of water and the number of people affected. Even here we felt the effects; much of our local industry centers on car components, and the temporary closing of automobile plants in Thailand caused a work slow-down. Mega-floods also brought chaos to Pakistan, the Philippines, China, India, Mexico, Japan (after the tsunami), and numerous other places.

In the United States, at least, 2012 has been the opposite: very dry. At the end of July, more than half of all land in the lower forty-eight states was in moderate to severe drought. Diminished grain crops were expected to push global food prices upward and cause inflation. According to an August 31, 2012, article in *Businessweek*, agricultural losses would exceed $10 billion and homeowners will likely spend more than $1 billion repairing cracked basements, foundations, and walls due to dry, shifting soil. The fire season has been quite intense; as the Billings, Montana, newspaper put it, "one for the books."

A September 2012 Oxfam report ("Extreme Weather, Extreme Prices") says that global changes in rainfall patterns and temperatures could lead to the cost of staple foods more than doubling over the next twenty years (from 2010 prices), with price spikes following periods of extreme weather. Sudden price jumps, such as those sparked by droughts or floods, are especially hard for vulnerable populations since there's no opportunity to plan or adjust.

These catastrophes are unpredictable and yet inexorable. We feel there's nothing we can do so we keep our heads down and pray that if a storm, drought, or fire hits, it hits elsewhere. Other than, say, clearing a fire path or stocking up on batteries, we're helpless. It's seen as the inevitable result of climate change, yet another unpleasant manifestation of the "new normal."

Allan Savory's belief that drought and floods are man-made and therefore not inevitable opens the way for a different response. With this principle in mind, Grasslands, LLC, seeks to apply Holistic Management to thousands of acres of land, creating islands of ecological resilience with regard to the water cycle. So that perhaps when, say, an inordinately heavy rain comes in the spring, the pasture can absorb the water and there's little runoff. Without big torrents streaming off the property, there's no rush of water to cut into the cultivated hills downstream—so there's no erosion. Since water isn't flowing so fast, it's able to soak in and turn the fields green. The good crop of grass that results feeds more animals. These animals enrich the soil and increase the diversity of plants, and this enables the land to retain more water and organic matter. Well-hydrated land rich in organic matter sustains a greater variety of microbial life. And when the next spring and summer see little rain, the water cycle is functioning well enough—in part through the "small water cycle," the evapo-transpiration of plants, that's enhanced when water stays in the ground—that there's no drought or lack of vegetation.

Replay this scenario again and again, using Holistic Management (with cattle, sheep, goats, horses) or other restorative models (agroforestry, pasture cropping, natural sequence farming), and those islands of resilience expand and connect and, in time, are no longer islands but rather large intact areas of revived ecology. Floods happen less frequently and droughts aren't as severe.

All summer I'd watch for news about the areas in Montana and South Dakota near the ranches. I checked for reports of rain that would change the grazing season's trajectory. They never came. The word *drought* began popping up in reports here and there, and by early July the lack of rain became a big national story. I saw that a fire had jumped the 212, the highway that passes near the Cinch Buckle Ranch; the area around Broadus was without power for a week.

At the very end of August I chase down the ranchers to see how they fared. I catch Brandon Dalton at the camper. He's in a chatty mood; the family has already moved up to Montana, so he's been on his own. Emmett is about to start kindergarten, so he and Brandy wanted the kids settled. The spread of land up there has four houses, each with five bedrooms, and several outbuildings. He laughs, observing that they're "going from one extreme to another."

He gives me a retrospective on the season, Holistic-Management-style: "Going into the season, the planning process forces you to really evaluate your resource—what your grass is, what you anticipate—and plan for the worst. In April and May, we had a lot of cattle lined up to come onto the ranches. We assessed our situation and asked: What can we do? We knew we had poor growing conditions. If we took the cattle on that would leave us praying for rain, which usually doesn't turn out very well. We made the difficult decision to not take twenty-seven hundred yearlings. We felt we did not have the conditions to. Instead, we took only eight hundred cow–calf pairs. That meant well over $100,000 worth of cattle that we didn't take. This was early in the year when no one was talking drought. In fact, ranchers were expecting a banner year because of the anticipated record grain crop. At that exact time we were thinking, *This looks really bad.*"

As it turned out, says Brandon, "we were pressed pretty hard with eight hundred cow–calf pairs. We'd decided we would ration out the grass we had from the get-go, knowing that if we got a lot of rain we could replan. It's gone about exactly as planned. By far the most difficult challenge we've had to deal with is water. We have enough grass to carry the cattle, but our water is lacking in a lot of places."

In terms of rain, he says, "it's been a tale of two ranches. At Horse Creek there's been almost no rain. We've gotten a tenth or quarter inch here or there, but up to a half inch doesn't go deep into the soil." The BR saw one mid-May thunderstorm "that really sunk into the soil and kept the ranch moist into June." In late July, he says, the BR had an inch and a half of rain two days in a row, while Horse Creek, twenty miles away, got nothing. "It's such wide-open country, I could see the storm brewing and watch it go just south of the ranch. Maybe we got a quarter inch. That's just how it is out here—a lot of thunderstorms that are really spotty. Wherever they happen to hit, that's where it is."

In Montana, they'll be hiring another full-time person, which could be a single person or someone with a family. They're considering a training program because of the extra available housing. "It's definitely not for everyone," says Brandon. "The closest neighbors are ten miles away, and you'd have to be prepared for that type of social environment." There are those who thrive in such a setting, he says. "Some people who grow up on ranches feel lonely going to cities. Here, you know everybody. In a city there's a lot of people but nobody says anything to you, waves at you."

Meanwhile, at the Cinch Buckle, Ron says, "we're a little more fortunate. There was a band of moisture that we got, several inches of rain in July, though still way below the average." The challenge has not been grass but the cattle's drinking water. "A lot of our water comes from man-made impoundments and reservoirs. What's happened is the water level has dropped and the sulfate level has climbed higher and higher."

This, he says, is an ongoing situation in the region, particularly in dry years. "If you force cattle to drink that water, you'll kill them. We started testing water because we were getting some inklings of a problem. In June we lost four cows that were watering in one spot right on the borderline of what is safe and not safe. I had sent in some water samples before the cows died and the results hadn't come in yet. When we went out to move them we found them dead." He wonders: Could he have saved them had they gone to move cows a day earlier? He's now testing continually, he says.

The difficulty is that Holistic Planned Grazing works best when cattle graze smaller areas intensively for brief periods. With many spots lacking drinkable water, they haven't been able to do this as much as Goddard would have liked. To address the water problem they're extending underground pipelines, which, Goddard says, runs about

$15,000 a mile. In conventional ranching cattle have a larger range, so managers wouldn't be faced with same water problems.

I ask about the calves that Kathleen had been feeding, and Ron tells me the last orphan calf joined the herd a few days before. He says that due to the timing it's unlikely the mother was among the seemingly doomed Texas cows. "Sometimes cows have twins, and as they're not very smart they can't keep track of both calves," he says. "In this case we couldn't find a cow so we took him in. You know, a cow can't even count to two."

Agreed, cows aren't very smart. But as the land has shown, in the aggregate—that is, in their behavior as a herd—they evince a kind of genius.

Torrential Rains

Fifth-generation rancher Zachary Jones, Grasslands, LLC's, division manager, whom I mentioned in chapter 1, was smack in the middle of the 2011 floods. Yet Twodot Land and Livestock, near Harlowton in central Montana, suffered few losses. I spoke with Jones in summer 2011, when the torrential rains that came fast after record snowmelt were fresh in his mind.

"The American Fork Creek, which goes through our land, turned into a large river this spring, which contributed to the Musselshell [River] being thirteen times its usual runoff. The creek has never had this kind of flow before," he told me. (The American Fork Creek runs into the Musselshell, a tributary to the Missouri River.) "We didn't have much water running off our pastures. Where it did run, it was clear—no runoff of soil." This is because Twodot has porous soil, good ground cover, and plants with deep root systems, all of which retain water and minimize soil erosion. Some of his neighbors, however, didn't fare so well. "Some lost entire diesel generators that run irrigation systems. Worth $20,000 to $40,000—those just washed away. Others lost barns or corral systems. Many lost livestock in barns or corrals. A lot of sheep drowned."

That Twodot was relatively unscathed Jones attributes to the fact that, thanks to his father being an early adopter, it's been administered under Holistic Management since the 1980s. He says years of Holistic Planned Grazing meant "increased carbon content of the soil, which holds way more water. It's like a bigger sponge.

Plus, increased ground cover to slow the water down so that it can soak in." When water soaks in it feeds plant roots, fills aquifers, and recharges springs; moisture is released gradually and over time. Increasing soil carbon means more water can be absorbed by the land. "If everyone could increase soil carbon 1 percent, it would have taken even a more extreme weather event" to cause the kind of trouble they saw that year.

Jones acknowledges that, ecologically speaking, land knows no boundaries—and that, he says, is humbling, and a reminder of the ranch's vulnerability: "We manage our twenty-four thousand acres pretty well, but there's hundreds and thousands of acres up the creek from us. That has a lot more to do with whether we have a tragedy with an extreme weather event."

In the northern Great Plains, where land is not badly desertified, you can increase the effectiveness of the water cycle in two to four years, says Jones: "Basically, you increase mulch—grass, litter, and manure—so that the soil is shaded and grass is not oxidizing. This significantly changes the soil surface to slow the flow of water. You also get better functioning of microorganisms in the soil and better cycling of minerals."

Jones worries not just about the loss of property and infrastructure, but about prospects for the future. "Thousands of bridges and roads have been washed out across the country," he says. "Everyone thinks they need to be replaced. But if you do a holistic plan you can look at your map and community and say, 'What do you really need to withstand events, where do the winds blow, how many roads and bridges do we need?' Look at the whole—don't rebuild where it's going to flood again. Rather, build in a way that you're not afraid of the natural ebb and flow of disaster and abundance. What just happened—nature doesn't view it as a tragedy, it's just the way it is. We need to do a better job of managing our land or else we're going to wash away.

"Slaps in the face like this provide opportunities for change. A drought definitely gets people thinking about how they can manage resources better. It's easy to see then that our ranch has more grass and more animals, and makes more money. A flood is a harder link for people to make. People think they have no control over it. We actually do."

Chapter Nine
The Soil Standard

> The stark fact that appears now, and which wrote itself across the Roman Empire, is that debt and taxation increase as the soil declines.
>
> —G. T. Wrench, from *Reconstruction by Way of the Soil,* 1936

> *All New Wealth Comes From the Soil*
>
> —title of a booklet by Carl H. Wilken, 1957

WAY BACK AT THE BEGINNING I'D MENTIONED that my interest in soil and its place in our lives was sparked by reporting about economics. During the 2008 economic downturn, it had begun to dawn on me that the economic system that framed our lives simply didn't make sense. If huge sums of money—as in *trillions*—could be conjured up or disappear at the behest of a blip on a screen or the stroke of a pen, how could this reflect actual wealth? I started asking the question, *What is money?*—an inquiry that led me down many journalistic pathways I never expected to take.

One thing that continued to bother me is the very basic cognitive dissonance inherent in our assumptions about money, the disconnect between economics and the natural world. Take, for instance, the notion of growth. In every policy analysis, news report, and campaign speech, the answer to all economic problems—poverty and unemployment, income disparity, the national debt—is to "get back on a path to economic growth." But there's a problem: Further growth (expanding industry, adding population, bringing first-world consumption habits to third-world communities) will ultimately diminish the resources on which to base wealth. In other words, there's nothing to grow on. The London-based New Economics Foundation (NEF) has an excellent report on this very topic called "Growth Isn't Working:

Why We Need a New Economic Model," the cover of which depicts a hamster. The paper begins:

> From birth to puberty a hamster doubles its weight each week. If, then, instead of levelling-off in maturity as animals do, the hamster continued to double its weight each week, on its first birthday we would be facing a nine billion tonne hamster. If it kept eating at the same ratio of food to body weight, by then its daily intake would be greater than the total, annual amount of maize produced worldwide. There is a reason that in nature things do not grow indefinitely.

Other NEF publications specify the point in the year at which the world breaches the limit of sustainable consumption, or, as they put it, the time at which we go into "ecological debt." For example: If European Union residents ate only fish from European waters, as of July 6, 2012, stocks would be depleted and the continent would become "fish dependent" on distant seas. And September 27, 2011, marked the day "humanity exhaust[ed] nature's budget for the year." It's been widely noted that if everyone in the world consumed resources at the level that Americans do, we would need several planets to sustain us, at least four or five.

We are all hamsters now.

While conventional economics is presented as "scientific," as a set of universals, it's actually based on a set of principles that made sense under particular circumstances, namely cheap energy, geographic expansion, rising population, and easily procurable natural resources. Barring these conditions, so as to maintain economic growth we've had to resort to some canny tricks, among them war (which kick-starts lots of economic activity), easy credit (low mortgage rates kept numerous industries bustling for a while), and financialization (folks have gotten very creative with carving up, trading, and betting on securities). Today the financial sector occupies a larger and still-increasing part of the overall economy; at this point more "wealth" is created by lending, trading, repackaging, and projecting funds than by producing goods and services. In today's hyper-securitized financial environment we end up with abstraction built upon abstraction, so that an "investment" may be several degrees of separation from the entity (product, business, intellectual property) we are allegedly investing in. This is how the entire financial system can flirt with collapse and no one can explain why.

We're now in a situation where we're bumping up against the limits of what nature can freely give us, and the strategies we've used to sustain growth either don't work (low-interest credit? ask those who've faced foreclosure) or are unpalatable to the public (we can only be involved in so many wars). The financialization of the economy continues apace—thanks to Wall Street's impunity—but one wonders how long this can last. In part because more people are on to them (witness: the brisk mobilization of the Occupy Movement), and because the system isn't working for many people (the huge numbers of unemployed, people who would like to retire, students and would-be students, et cetera). And yet growth remains the only prescription our experts can offer. As the late Kenneth Boulding said of his own profession, "Anyone who believes exponential growth can go on forever in a finite world is either a madman or an economist."

Perhaps the most revealing example of the skewed relationship between economics and the environment is the use of the word *externality*. This crafty term means that any environmental side effects that ensue in the course of business are essentially written off. Instead of companies taking a financial hit from any ecological damage they impose, they're allowed to say "Sorry, not my problem" and walk away. If effluvia from Factory X pollutes the air, water, and soil in the area, the cost is borne by those who breathe that air and live downstream—the public—rather than the principals or stockholders of Factory X. This means that a huge amount of economic activity—the costs, say, of higher asthma rates or additional water treatment—is not accounted for in anyone's business plan. It's kind of an institutionalized cooking-of-the-books, or at least the leaving out of some important ingredients. As a result we get two parallel balance sheets moving in different directions: X Company looks better and better as profits rise while community expenses and liabilities mount. How can these two ledgers exist within the same reality? They can't, really. But this fiction will persist until we correct that accounting.

While this venture into soil may have been sparked by economics, it's taken numerous detours into, among other topics: the flow of carbon and water, the heroic behavior of fungi and worms, the missing minerals in our food, the enigmatic notion that a small number of cattle can harm soil while a large herd can heal it. Right now the theoretical musings over markets and policy feel far away. Yet soil, as we've seen, is intrinsic to many natural cycles, as well as to the production of food. Certainly, this venture into soil has much to bear on our understanding of wealth and commerce; it's a matter of making the connections.

I'll divide this exploration of soil and economics into two parts, one practical and one meta.

The practical comes down to this: A focus on soil health enhances production, and therefore earnings; it saves money, by eliminating expenses that problems with soil would cause; and it serves as an investment in the future by improving conditions for agriculture and infrastructure.

When it comes to practicality—and economy—North Dakota is a fine place to start. This largely rural, sparsely populated state is, incidentally, the only one to have enjoyed a large budget surplus over the last several years. This is due to several factors, including a robust agricultural economy and new energy production, but also the Bank of North Dakota, a public bank established nearly a century ago that keeps money in the state, drawing on public wealth to provide credit to citizens and local enterprises.

Jay Fuhrer, district conservationist with the Natural Resources Conservation Service in Burleigh County, also, perhaps jokingly, attributes this relative flushness to the habits and character of his fellow North Dakotans. The state is prospering economically, he says—"There's no unemployment here, unless you choose not to work"—but the cultural norm is not to show it off. "People [are] living just the way they always have. Here the guy you see in a flannel shirt driving a very old Chevy is probably a millionaire," he says. "And it's likely he's collecting cans for recycling in the back of his truck."

We're back in Bismarck, in the bright, uncluttered office of the Burleigh County Soil Conservation District (BCSCD). Jay, my guide to the area whom we met in chapter 6, has been giving me a glimpse into the

North Dakota way of life. He has short gray hair and a tidy mustache and wears black jeans and a red shirt with the NRCS logo. There's something deeply sane about him, a reserved, measured quality; he has a slow-release wit and a wry, colorful way of speaking, as when he refers to moving cattle out of feedlots as "giving the cows back their legs." He rarely travels out of the region and says he feels little need to see any big cities, such as, say, New York. He and his wife have three grown children who live in the Bismarck area. North Dakotans don't put much stock in glitter and glitz and the trappings of wealth, he says, quoting a rancher who once told him, "If I can't stand on it or eat it, I don't want it."

North Dakota was one of the last states to be settled by Europeans, Jay tells me. Many pioneers were Germans or Norwegians lured by the Homestead Act: the chance for their "little piece of heaven," 160 acres of untrammeled northern prairie, to make of what they could. One significant immigrant group—and Jay's through his father's side—is "Germans from Russia." These are Germans who settled in Russia, mostly near the Black Sea in what is now Ukraine, when Catherine the Great, herself of German descent, made it attractive to do so. By the late nineteenth century those special privileges—notably the exemption from serving in the Russian army—were rescinded, leaving these choices: become Russianized, be shipped to Siberia, or leave.

"My grandmother was born on the ship coming over. I still have the homestead document," says Jay. "The first winter was the hardest. People slept under their wagons or built sod houses. Others ordered houses from Sears, which were shipped out of Chicago by rail. People would take wagons pulled by horses or mules to get them and then had to put them together. There was a lot of self-sufficiency. Everyone was a farmer, butcher, carpenter. I remember as a kid we still butchered in the fall. Neighbors would get together to help butcher and make sausages."

His father spoke German and his mother spoke Dutch in addition to English. "I went two miles to the one-room schoolhouse every day," he says. "There was no town. It was just a building in the middle of the prairie. The teacher would stay at the houses of the students. There was a potbellied stove, and a barn where you could put your horse."

Area schools consolidated in the 1960s, but Jay has "nothing but good memories of the one-room school." He attended high school in Strasburg, the home of Lawrence Welk. "He'd come to school once or twice a year and play polkas and dance," he recalls.

The kind of small, diversified farm Jay grew up on has little chance of survival in the current economic climate, he says. "Someone can absorb another section of land and not even notice it. Here in North Dakota, the land tells you you'll never get a tractor or combine big enough. That's been the trend for so long."

Jay studied agricultural economics at North Dakota State, and since 1980 has been with the Natural Resources Conservation Service (previously called the Soil Conservation Service). The NRCS, part of the US Department of Agriculture, was established in the 1930s in the bleak aftermath of the Dust Bowl to turn around our deteriorating farmland. In a bulletin to state governors, President Franklin D. Roosevelt famously wrote of the need "to conserve the soil as our basic asset. The Nation that destroys its soil destroys itself." Incorporated into the era's soil-related legislation was the mandate to form soil conservation districts; this requirement remains, though many are linked to other departments and may have different names and functions. To Jay, these districts represent opportunities to bring the importance of soil health to the fore, to go beyond providing basic services to putting soil health and function at the center of regional environmental and economic enhancement. But before he could start making that happen in Burleigh County, he had to push past his own assumptions and training.

Early on, one task was to help farmers work with the Highly Erodible Land Compliance requirements. "They needed to make conservation plans to be eligible for program payments," Jay says. "It was a real uphill battle, to figure out how much soil loss you could have and still be eligible." That was the standard approach, he says, "rather than repairing it to be stronger than before." Over time, he says, he encouraged people to "move away from how much loss, and said, 'Let's not even look at the loss. Let's have *no* soil loss.' We were starting to see that farms with residue on the surface, minimal soil disturbance and high crop diversity had essentially eliminated soil erosion, which then resulted in improved soil health." So he knew it was possible, and had an inkling of what it would take.

"The first half of my career I totally wasted," he says. "I was busy treating symptoms. I hadn't seen 'the matrix' yet. You start to see a few soil health dots. When you begin to connect the dots, an interesting thing happens: You see so many other dots that come into focus." Seeing the folly of rewarding farmers for a reduced loss rather than for soil health improvements "was the aha moment for me," he says. "What to a farmer has more worth than the land? When you come to a farmer with mineral support, biodiversity, ways to make it better, it's a whole different thing."

Another district program involved putting in sod waterways to prevent the formation of gullies. "Once you had erosion starting and you couldn't use farm equipment, we would survey it, a contractor would build it and seed it to grass," he says. "In retrospect, I see this as a symptom of problems with the soil—the water wasn't getting into the profile. If we had focused on soil health to begin with, we probably wouldn't have needed to intervene. One day we decided we needed to change. We focused on the holism of nature between livestock and crops and pollinators and cover crops. In the last ten years we've done exactly one waterway. Each one of these costs thousands of dollars. I look back and think of the money we could have saved. But I couldn't see the problem at the time, only the symptom."

Soil Equity

Jay has it all planned out for me: Over the next two days he'll take me to four farms/ranches, all regular stops on BCSCD's annual Soil Health Tour, so I'll see a range of landscapes and approaches to improving soil as a basis for a profitable operation. But first: a simple soil slake and infiltration demonstration. It isn't enough to take notes; I should understand how soil works. Usually, he says, he sets this up on the gate of a pickup somewhere in a field. But, well, we were here.

Our experiment: Compare two cropland soils. In one, the land had high disturbance (tillage), low crop diversity, no cover crops, and no livestock; the other offers low soil disturbance, high crop diversity, cover crops, and livestock. The first demonstration is a slake test, and with the help of a "rain simulator" (a sour cream container punched with holes) we watch the water's effect. After several moments the difference is evident: The high-disturbance soil breaks apart ("not enough glomalin or 'glue,'" says Jay) and the water clouds up. The second part is the infiltration test, and when we check back a while later the high-disturbance soil has water ponded on the soil surface. In the undisturbed sample, water has slowly infiltrated through the soil and into the aluminum collecting pan below. Functioning soil has pore spaces that allow for water to move through. Without pore spaces, says Jay, "water sits on top and the farmer says 'we gotta do something.'"

Next we measure nitrate leaching. Jay swipes nitrate strips through the distilled water as it infiltrates through the two soils. Both samples show evidence of nitrates, with the high-disturbance soil showing a higher release.

"It's been our observation that as you tend toward a monoculture, your input costs go up and soil problems go up, too. As you move toward biodiversity, the input costs go down and symptoms go down. A monoculture grown every year with high soil disturbance reduces the role of the soil to just holding the plant upright."

It seems so clear that it's the wrong way to go. I ask, "Well, then why does everyone do monocultures?"

"There's a natural human tendency to lean toward simplicity, which here means one crop," he says. "When you're profitable, you have a reluctance to move away from it. You'll ride that camel a little further. And when that camel drops dead in the desert you're going to kick him. Hard."

The following day, Saturday, is overcast. It rained the night before and the air still has that heavy, laden feeling. We take the NRCS truck out to Menoken, about twenty miles from Bismarck, to Richter Farms, where Marlyn Richter, a burly man with blue eyes and a red baseball cap, greets us and says to me, "You want a pop?" He ushers us into the office, a large room in a barn with a long table and a row of green John Deere toy replicas lining a high shelf. "Nice rain, uh, Jay?" he says.

"Million-dollar rain. Real gentle too." One of fourteen siblings, Marlyn co-owns and runs the dairy/beef/grain operation with his brother Patrick and their parents. He's also of Germans-from-Russia descent. He and Jay remark on the shadow side of the industrious legacy they share. "I feel guilty if I'm not working," he says.

"Every day at five I feel guilty because I grew up milking," Jay counters. The two shake their heads.

The first change in the right direction, Marlyn tells me, was shifting to no-till farming in 2000. At least a third of US agricultural land is no-till. Often this is accompanied by heavier reliance on herbicides, but this needn't be the case. Through using cover crops the Richters, for example, cut herbicide cost by half. "Especially with our sandy soils we needed to do something," he says. "It tends toward erosion. Our sandy soils have low water-holding capacity. That's the hand that's been dealt." The switch had several results. For one, using the tractor less meant significant fuel savings. Also, the land was spared the damage caused by tilling: turning the soil undermines ecological processes—all those microbes and fungi doing barter around the root zone—and disturbs pore spaces so as to impede the water cycle. Today, he says, water that does run off is clear, like the "good" soil in our demo, suggesting minimal erosion or chemical runoff.

The next change was adding a cover crop mix to create crop diversity, add nutrients, and build soil organic matter. In terms of day-to-day farming, cover cropping saves about $20 an acre in growing costs, Marlyn says. This mantle of green means that plants are working year-round on the land, harvesting sunlight and ferrying carbon down into the soil. "You can't keep writing checks without putting anything in," he says. "The homesteaders could get away with it for years. Soil organic matter was high then. Now we're farmed out." Over the last decade, soil organic matter on the farm has increased 2 percent or more.

The Richters also kept more residue—"armor" in Jay's lexicon—on the soil, strategically moving the range cattle so they wouldn't eat plants down to the ground. The grasses and forbs feed livestock while, below the ground, their roots give off exudates to feed the soil biology. Soil armor slows flowing water and stops erosion, and is pivotal to maintaining soil temperature (see the chart).

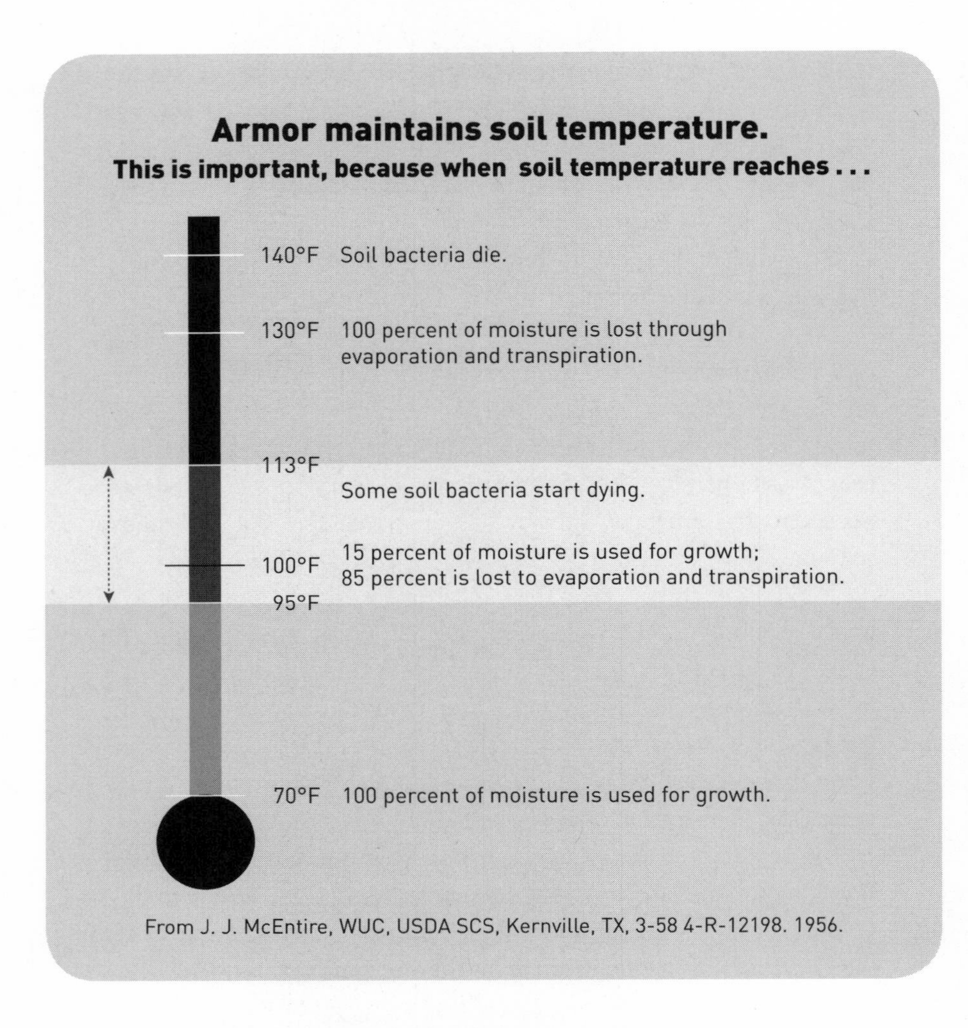

Soil management has helped the Richters "deal with the peaks and valleys" of farming, Marlyn says. "We've increased our production by 40 percent by growing grain corn when we moved into the no-till cropping system with high crop diversity and armor on the surface. I used to think, the straighter the row, the better the farmer. Now the messier it looks, the better I feel."

We take a long, muddy dirt road to our next stop, Black Leg Ranch in McKenzie. Jay tells me our host, Jerry Doan, is "English by way of

Canada." So when Jerry, a tall fellow with glasses, side-swept gray hair, and a hearty smile, strides out to us and starts chitchatting with Jay, I'm surprised by the ringing American accent. The Doans, I learn, came from England on the *Mayflower*. Family members migrated to Pennsylvania, where they generally wreaked havoc and became notorious as the "Plumstead Boys"; in 1783 the Pennsylvania General Assembly declared the Doans "robbers, felons, burglars and traitors." The clan fled to Canada until 1882, when one George Doan ventured back and bought the land that would become Black Leg Ranch, so named because it was the first ranch in the area to feature Black Angus cattle. A century-plus later, the ten-thousand-acre ranch combines custom grazing and an agritourism business called Rolling Plains Adventures.

Jerry shows us one of the hunting lodges, an airy, cedar log house, which his two older sons and their wife and fiancée, respectively, are readying for guests. The main house, also cedar log, is filled with belt-buckle trophies (his youngest son is on the Montana State University Rodeo Team, and his daughter was Miss Rodeo North Dakota and third runner-up for Miss Rodeo America), impressive displays of taxidermy, and portraits of handsome youths with great teeth adorned in cowboy paraphernalia. Jerry has used Holistic Planned Grazing for a number of years, and says he's seen "immense change" in the land.

"We're very sandy," he says. "When I was a kid we had lots of blowouts. There's still a spot here or there but a lot of that has been covered. The diversity has grown immensely. You can see it in the colors of all the blossoms. The diversity really attracts the wildlife." For Jerry, wildlife diversity speaks directly to his business, as people sign onto Rolling Plains Adventures to fish and view and hunt pheasant, waterfowl, coyote, and deer. (During a typical deer hunt, your odds of seeing a buck of 100 to 150 inches: 95 percent.) "We've got more wildlife than ever and more livestock than ever," he says, attributing this to cover crop mixing, more litter, and biodiversity starting at the level of soil. "You can adjust cover crops depending on what you want. For example, enhancing flowering species for pollinators."

A modern rancher's biggest expense, says Jerry, is the cost of feed during the winter. "We want to stop 'feeding' our cattle. We want to keep the cattle on the land," he continues. "It never made sense to haul hay in and haul dung out. This year we made it all the way to March 1 on native range and cover crops. The savings was over $50,000 compared to our former system of feeding hay all winter. The savings could be a salary or family living as we add that next generation to the business. And we might be able to double that number. Aside from what this does to the soil health and water quality, including the creek, which is important for hunting."

Sunday morning we go to see Gabe Brown on the outskirts of Bismarck. The weather has changed, and now that it's sunny I sense the great horizontal boundlessness of the plains, a vastness that had been constricted by cloud. "We're going to harvest sunlight today," Jay says, looking up to the sky. "You can almost feel the cleanness, that everything's green, as if it's healing."

He and I wait in a field while Gabe talks on a cell phone nearby. "It's a normal pose for him," Jay says with mock annoyance. "He's on the phone all day." Gabe has become something of a celebrity in sustainable farming circles, especially since he was awarded a 2012 Growing Green Award from the Natural Resources Defense Council. In paying Gabe a visit, I am but one of more than a thousand a year who stop by. Jay digs into the earth with his spade—he's rarely without it—and pulls up some soil. He rubs it with his fingers and lifts it to his nose. "This is prime soil," he says. "It breaks into individual granules but you don't get your hands dirty. It's probably how soil used to be. We're so used to degraded soil." The "good" soil in our infiltration experiment came from Gabe's farm, he says.

And the "not-good" soil?

Nowhere I'd know, he assures me.

Gabe saunters down the hill and pockets his cell phone, despite which it will ring multiple times over the next half hour. "Beautiful morning. Can't beat it," he says. Gabe is solid, broad in the jaw, and genial. I notice that when he refers to soil, he often uses the word *resource*. The farm, he tells me, is fifty-four hundred acres, "a little on the smaller side for the area. It's getting smaller all the time. As we focus on the resource, we've found we need less acreage."

Gabe bought the land from his in-laws in 1991, went no-till, and immediately hit a bad patch. Between 1995 and 1998 the farm lost four crop years in a row due to hail and drought, which put the family at the financial edge. "I couldn't afford to buy inputs anymore, and I had to do something, had to figure out ways to provide nitrogen. I came to realize it's all about the soil, and that you don't need commercial inputs if you've got healthy soil. It's about focusing on the resource with rotations, cover crops, and I really believe livestock need to be there, which is difficult because the management level needs to be higher."

With its array of inputs, the agricultural industry offers means of "treating symptoms, not solving problems," he says. "Usually, any problem I see out there I can address. If you ask me, why do people need to use commercial fertility? That's easy: The soil's in poor health. Here we no longer use fungicide on our long-term no-till. We no longer need chemical fertilizer and we haven't used pesticides in twelve years. We still use herbicide when seeding, though we've brought it down 75 percent. When you're switching a cropland field that's been conventionally farmed to no-till, you can't just cut it all out. You have to build it first."

Gabe looks down at the ground and picks up a daikon radish nearly the size of a baseball bat. "Every crop we grow here has a purpose. This here improves infiltration and is a nitrogen storage tank. Radish will scavenge nitrogen and, when it breaks down, release it." Here and there are turnips from the last season. Jay points out the first spinach-y potato leaves popping up.

As for finances, Gabe says that "it takes an average of twenty-one gallons of diesel fuel to plant, grow, and harvest an acre of corn. Here we're doing it in five. If we can save 75 percent of our fossil fuel bills, we're doing well." While he's looking to "shrink" the farm ("so we can manage it in a better way"), the trend toward bigger operations continues. "A lot of that is driven by economics. But I will disagree—economical does not have to mean big. We produce a bushel of corn for $1.21. Typically, it's $3-plus. If you're in the mindset of conventional agriculture, you're still using the inputs." Which means paying for them, which gets more expensive as oil prices rise. And dealing with the consequences in the soil.

Gabe doesn't see change coming from the large producers—the agribusinesses. "The smaller operators are more able to 'get it.' The large ones are the last man standing." Soil improvements are a kind of equity, he says. "At today's fertilizer prices, each 1 percent of soil organic matter contains $650 per acre worth of nitrogen, phosphorus, potash, sulphur and carbon," he told *The Furrow*, the monthly magazine published by John Deere in 2011. Since soil organic matter has gone from 2 percent to 4 percent, "this means . . . we have $2,600 per acre worth of those nutrients locked in the top six inches of soil. The trick, of course, is to make them available to plants, and that's where spurring the soil's biological activity comes into play. Instead of focusing on feeding the crop (with commercial fertilizer) we're focusing on feeding our soil so it feeds the crop."

We need to think differently about economics, he says. "We can improve the economy without growth. We can do it without more bushels. I get fed up when I hear, 'We've got to increase production to feed the world.' What's the good of increasing production if it's not healthy? Let's look at wealth in the context of the human health crisis. If it's a healthier product we're growing, we're lowering costs."

The impact of weather-related problems shows the importance of soil health, he says. "Today, if there's a drought, will I suffer? Some, but not as much as others. There's an economic ripple effect: Every flood or drought adds costs down the line." This includes, he says, the web of subsidies, incentives, and insurance programs intended to minimize the risk to growers. "They say farmers need a safety net. I think the soil should be our safety net. If you add up the dollar figure for improved health, carbon sequestration, lower fossil fuel bills, and resistance to weather extremes, it's a lot. The real passion for us is: What are we leaving for the next generation? I don't like the term *sustainable*. I don't want the land to stay degraded. I like the word *regenerative*—we've got to regenerate the resource."

The last site is Menoken Farm, 150 acres purchased in 2009 by the Burleigh County Soil Conservation District as a demonstration farm for soil health programs. We exchange the (by now well-mud-spattered)

white van for a small utility vehicle and Jay takes me out into the fields. From the bounce in his step I can tell this place is special to him, that this is where he gets to try out his own ideas, not just offer supporting advice. "That darned Gabe!" he says as he pulls a sharp turn onto a narrow greenway between crop rows. "Always gotta get a potato up ahead of me."

The fields are richly green with some gold from old growth; mostly I'm looking at cover crop. But Jay's eye is trained to see the potential. "We're experimenting with cover crop mixtures," he says. "We have ten fields. One has commercial fertilizer. That's our control field. The first year we didn't use fertilizer I got withdrawal. Because I grew up with it." Rather than fertilizer, they amend with compost and compost tea (a biologically rich concentrated liquid derived from compost), and have sheep and cattle graze at different points in the year.

When the vehicle stops the air is calm but busy, full of birds and bird sounds, butterflies and insects. "We're measuring the complexity effect," he says. "On each field, we add a cover crop. I'd like to add a biannual. That way you can plant in the spring and not have it go to seed." For the 2011 season, the farm was part of a program through the Hunger-Free North Dakota Garden Project, a community effort that yielded a bounty of produce, including fifteen hundred pounds of potatoes, for local food banks. Menoken Farm has been getting high bushel-per-acre rates, particularly where full-year cover crops were used. But yield alone isn't enough for Jay. "I want to start taking measures of the quality of the food," he says.

Back when I was planning my trip, Jay had warned me that spring wasn't the best time to see North Dakota farms—there's much more to see in the fall. Unfortunately, I could only go in the spring. Later in the season I look for Menoken Farm on the Internet and find a picture of Jay surrounded by vivid green corn plants, dwarfed by their height and abundance. In a video he tells the camera, "This corn is run totally off the energy of last year's cover crop." A photo from last year's Soil Health Tour shows a group of people standing in one of the fields. You can hardly see the men, the plants are so high: corn, sunflower, something with a purple flower. I was beginning to understand: soil health, soil biology, plant diversity, resilience to weather fluctuations, fertility, quality, and economic vitality all go together. As Jay would put it, I was beginning to see "the matrix."

Back to economics. Upon posing for myself that deceptively simple question, *What is money?*, one topic I started writing about was alternative forms of currency. Most of us accept that money is that green paper we carry in our wallets, and that this arrangement—paper bills keyed into a centralized monetary system—is sacrosanct. However, currencies can take many forms. Nor is there any reason to stick with one currency. In fact, you could argue that dependence on one currency is like total dependence on one crop: There's always the risk that crop failure or a cutoff in supply will topple the whole system. Out in the world today you'll find all sorts of currency experiments, from local currencies like the Brixton Pound in London and Berkshare Dollars in the southern Berkshires in Massachusetts to Time Bank exchanges based on the hour to computer-driven setups like BitCoin. There are theoretical reasons behind various micro-currencies, such as keeping wealth in the community and enhancing the velocity of money—an example of this is the Chiemgauer Regional Currency in Germany, which incorporates a penalty for hoarding so that money continues to circulate. A local currency can also provide feedback to suggest weaknesses or imbalances in the economy, a function that a broad-scale fiat currency like the dollar cannot perform with the same regional precision. (We see how this has played out with the euro.)

For me, it was the very notion that you could create your own currency that threw it all wide open. If the money we organize much of our lives around is not, as I had thought, inviolable, then what is the nature of wealth? Why should we accept the inequities inherent in this particular system, in which monetary wealth flows to the money centers regardless of who provides the material or performs the labor? Might there be better ways to store, measure, or exchange value? These were some awfully big questions to sit with on my own, so I linked up with the E. F. Schumacher Society (now the Schumacher Center for a New Economics) nearby, and learned of others grappling with these themes. The field of New Economics, which sees the purpose of the economy as serving people and the environment, provided a frame for inquiry. In conventional economics, too often people and the environment seem to be regarded as subsidiaries of the economy. Economically speaking, then, people are expendable and interchangeable, accorded value to the extent that they contribute economically. And the environment is a storehouse of potential wealth, wealth that's dormant until it's removed or developed.

Which brings us back to that disconnect between the economy and the natural world, a fantasy that we continue to believe at our peril. "The only true economies are nature's ecosystems," says Wes Jackson, founder and president of The Land Institute in Salina, Kansas, which is exploring agricultural models in harmony with ecological systems, with an emphasis on perennial grains. He means "economy" in the sense of "thrifty management of resources," for this is where ecological systems excel: self-sustaining and self-regulating, responsive to change, nothing is wasted (in contrast with our industrial, capitalist economy, where waste is rife and unchecked). The decidedly unnatural ecology of industrial agriculture is often justified in terms of "economies of scale." Yet as we've seen, the practices that achieve scale—monocultures, high inputs, feedlots—are ecologically unsound; the benefits of the system, therefore, are ultimately an economic fiction. Whether or not we choose to acknowledge it, the economies that we create are embedded in the natural economy.

The Soil-Economy Matrix

When you come down to it, money is a metaphor—actually, a metaphor many times over. We accept that money isn't in itself wealth, but its symbolic stand-in: The twenty you hand your teenage son or the deposit that shows up electronically in your checking account represents a tiny fraction of the vast pool of national wealth. (Before 1971, when we ditched the gold standard, this would have meant an actual, albeit infinitesimal, quantity of shiny metal.) On an emotional level money is metaphor in that it inevitably carries some personal meaning—security, nurturing, temptation. The word *metaphor* comes from the Greek *metapherein*, which means "to transfer." Ultimately this is the essence of money. If you can't transfer with it, if in some way it fails to yield up something you want or need, that piece of crinkly green paper or mark on a screen isn't worth a thing.

Which raises the question, metaphor for *what*?

As it turns out, money and wealth have often been described in terms of *soil*. In *An Agricultural Testament*, Sir Albert Howard talks of industrial fertilizer as "transfer of the soil's capital to the current account." Agronomist William Albrecht wrote, "All the capital in all the banks cannot substitute for the soil of the land. We know of no bank with all its money that could by means of that wealth have a litter of pigs, lay an egg, or give birth to a calf." He alluded to the geopolitical implications of this natural capital, calling war a "result of the global struggle for soil fertility." To Albrecht, the ongoing work of nations, whether conducting war or building and sustaining an economy, consists of "mobilized soil fertility."

And here's a quote from Carlo Petrini, founder of Slow Food International, in his foreword to Woody Tasch's book *Inquiries into the Nature of Slow Money: Investing as if Food, Farms, and Fertility Mattered*: "At the base of the economy is soil fertility. If we use money like synthetic fertilizer, we will get artificial growth, which can only last for awhile, but which lacks sustaining relationships with the earth."

In *The Solutions Journal*, ecologist John Todd takes the analogy yet farther and proposes that carbon—specifically the carbon found in the soil—serve as a form of currency:

> Humanity has always been carbon based. The carbon that supported us through most of history was slow carbon embodied in trees, other plants, and animals. Since the Industrial Revolution we have shifted to using fast carbon in the form of oil and natural gas. Fast carbon is mainly finite and nonrenewable. What if we used carbon as a universal currency? What if people around the world were paid to capture and sequester carbon, particularly in soils? What if enterprises that emit carbon into the atmosphere, including, for example, coal-fired power plants, had to pay for the right to pollute based upon every ton of carbon they emitted? A tectonic shift in the way the world conducts its

> business, from farming to aerospace, might ensue. Let's continue the conversation. The stakes are too important not to.

He could have been channeling Christine Jones, who has said: "Carbon is the currency for most transactions within and between living things."

What "slow" and "fast" carbon have in common is that they are the products of photosynthesis: the conversion of the sun's energy into chemical energy, a process that fixes carbon. Our dominant energy sources (coal, gas, and petroleum) are essentially fossilized solar energy by way of the fixing of carbon. Right now our economic system depends on running through carbon rapidly without fixing an equivalent amount.

In other words, basically what we've got now is an oxidizing economy. We've been undoing the photosynthesis of the past, a process that releases heat and sends carbon dioxide into the atmosphere. From the standpoint of land, our oxidizing economy dries soil in such a way that plants and microorganisms cannot be sustained, creates conditions in which water doesn't infiltrate and aquifers are not recharged, and disrupts the climate. The other direction—photosynthesis—promotes plant growth, creates conditions for soil to absorb and hold water (thus building freshwater stores), supports microbial life, cools the air, moderates the climate, and sequesters carbon dioxide.

Remember our bumper sticker from the first chapter? OXIDIZE LESS, PHOTOSYNTHESIZE MORE. We can apply that to our economy as well as to our ecology. The present economy is ripping through the wealth of photosynthesis. We need, instead, to build that wealth—because that is the wealth upon which all other wealth depends. The ecological ideal would be a balance between oxidation and photosynthesis, but we're so off kilter that this is a long way off.

How feasible is it to shift in the direction of photosynthesis? This book is filled with examples of people who are making this happen by promoting ways to: build carbon stores in the soil; keep water on the land; stop and reverse desertification; focus on ecological systems rather than individual species; reduce chemical inputs; and maintain diversity in crops, native plants, and microbial life. And—we can't forget our cows—bring herbivores back onto the landscape.

In a 2005 talk at The Leopold Center for Sustainable Agriculture at Iowa State titled "The Farm as Natural Habitat," Laura Jackson, Wes Jackson's daughter and professor of biology at the University of Northern Iowa, said: "In most areas of the Upper Midwest, land in agricultural production is barren dirt for nine months of the year. Because of our corn/soybean rotation, we're looking at a system of collecting solar energy about three months of the year. The rest of the time the land has very little cover on it, very little green leafy cover to collect solar energy . . ."

Jackson included a slide that depicted, via satellite imaging, the "greenness index," or plant cover, over a period of two weeks in June. She said: "The maximum amount of solar energy comes to Iowa on or around June 21, and Figure 2 shows that a big chunk of the Corn Belt is virtually bare, brown to yellow, on the same days that solar energy is at its maximum. What a waste, right?"

This is where we can make the shift: on the expanses of land in our country and around the world that, photosynthesis-wise, are underperforming, or where, to use Peter Donovan's terminology, we've got improvident "sunshine spills." We can make the shift by reinvigorating our soils, which serve as a hub for so many of the ecological cycles that support life and sustain and build natural wealth.

We can proceed as we've been going, abiding by an economic structure divorced from the natural world that functions according to its own arbitrary and increasingly illusory rules. Or we can work to develop economic models that invite us to rejoin the ecological systems on which we depend. Because when we have a system in which wealth depends on processes that destroy natural capital, we're only kidding ourselves. As the late ecologist Howard T. Odum has said, "Inside the human system, money can expand exponentially, but 'real wealth' remains limited by energy, materials, and biophysical processes." In her talk, Laura Jackson cautioned us about "the vast veto power of nature over what we would like it to do." For in contrast with the flexibility inherent in our economy, nature is non-negotiable.

There are people thinking this way. One person who well articulates the link between natural and human economics is filmmaker John D.

Liu, who left network news in the mid-1990s to focus on environmental media and education. He writes:

> From the study of natural ecosystems comes an economic answer that goes to the fundamental question of 'what is wealth?'. Although everything that is produced and consumed comes from the bounty of the Earth, according to current economic thinking, the value of ecological function is zero. We now calculate the economy and money as the sum total of production and consumption of goods and services. By valuing products and services without recognising the ecological function from which they are derived, we have created a perverse incentive to degrade the Earth's ecosystems.

One of Liu's projects has been documenting the ongoing rehabilitation of the Loess Plateau in China, where coordinated projects have slowed the erosion of soil. One YouTube clip called "The Stupifyingly Simple Solutions to Preventing Drought and Flooding" (on the "whatifwechange" channel) draws on his observations:

> The problem has been that we've been looking at the soil only from the perspective of increasing productivity. If we look again at this with the goal of increasing ecological function we can employ millions in actively fighting against drought and flooding and simultaneously increase productivity. The key is that we all need to work together. We need to revegetate the degraded parts of the earth and employ ecological principles in our agriculture, industrial and urban areas. We need to realize that wealth is not coming from manufactured goods and from commerce. Wealth is coming from natural ecological function. If we understand this we can base our monetary systems on ecological function. And to do conservation of the earth will be to protect wealth. And to restore degraded areas will be to increase wealth.

The Slow Money movement, inspired by Woody Tasch's 2009 book, challenges an economic structure that's too fast (as much "wealth" consists of electronic blips shuttling from screen to screen in speculative transfers), is disengaged from the natural world, and fails to serve the needs of communities. The Slow Money Alliance has begun its "from the ground up" efforts with helping to "seed" small, locally based food enterprises; upward of $20 million has already been distributed this way. Inherent in Slow Money's critique of our current economy is that we've been heedless of a crucial generator of real wealth—the soil. From the web page articulating the group's vision: "The soil teaches us that we must put back as much as we take out to ensure long term health and a strong, secure, restorative economy . . . When we erode our soil, we erode our social capital, we erode community." One new funding vehicle, called the Soil Trust, "will pool a large number of small donations to create a permanent, philanthropic investment fund dedicated to small food enterprises and soil fertility."

Slow Money exemplifies what Peter Donovan refers to as "managing *for*," in this case managing for local economic vitality grounded in soil fertility. Right now, people are not economically rewarded for this kind of management. In the agricultural sector, for example, our subsidy structure is such that farmers and ranchers are essentially rewarded for mucking up the soil rather than building it. The emphasis on scale creates incentives to grow single commodity crops, such as corn, wheat, and soy. And yet soil is part of the commons—a resource that every single one of us has a stake in sustaining.

All this may seem a far cry from our strolls through the fields in North Dakota, encountering Marlyn with his John Deere miniatures, Jerry and his hunting lodge, Gabe with his nitrogen-scavenging giant radish, and Jay with his custom crop blends. But what people like Marlyn, Jerry, Gabe, and Jay are doing is "managing for" photosynthesis over oxidation in several ways: encouraging diversity, avoiding tillage, using cover crops, and generally improving the soil. Our agricultural and economic policies can reflect this. Not that we have to change all our currency to the "soil standard," but we should certainly take it into consideration.

I'll leave you with Kurt Vonnegut's iconic comment: "We could have saved the Earth but we were too damned cheap." But you see, it wouldn't be so expensive. It's just that up to now, we—collectively—have tried to pretend that we could proceed with an economic model in which we ignored earth's natural cycles, and the role in those cycles played by soil. We simply took our eyes off the books.

Acknowledgments

I'D LIKE TO THANK MY FAMILY AND FRIENDS for humoring and even sharing in my newfound interest in soil and its place in the world. I am extremely grateful for all my teachers in this realm: The journalistic term *source* is far too cold a word given the education, time, and generosity of spirit involved in cooperating with a writer who's learning on the job, and for the sense of partnership that develops. Their words, observations, and insights appear throughout the book and are what made it possible. I'm also grateful to my nephew, Julius Schwartz, who came all the way from Scotland to keep me company on my travels through the northern Great Plains. I want to give special thanks to literary agent Laura Gross, who embraced this book immediately, and to all the people at Chelsea Green Publishing, who have been unfailingly respectful, professional, and enthusiastic through the entire process.

Notes

1. Here it's important to differentiate legacy carbon—CO_2 that's been released through human activity, including farming and industry—from ambient carbon, naturally occurring CO_2 that's vital for biological processes, notably plant growth. These sources of CO_2 coexist (actually, commingle), so it's a distinction we need to maintain in our minds. Just as there's an upper limit to the CO_2 in the atmosphere that earth's systems can tolerate, there's a threshold below which plants and animals cannot survive.
2. But there is only one atom of carbon in each of the molecules. The twenty-five-times-more-potent figure is calculated by comparing the two gases on a kilogram-for-kilogram basis, not a molecule-for-molecule basis. The differing molecular weights mean that there are nearly three times as many molecules of methane—and, therefore, atoms of carbon—in a kilogram as there are molecules of carbon dioxide. The more accurate global warming potential of methane is around eight or nine times that of carbon dioxide, and even this figure depends on some further assumptions.
3. Yeomans, P. A., *The City Forest: The Keyline Plan for the Human Environment Revolution*, Keyline Publishing Pty., Sydney (1971). Thanks to Abe Collins for alerting me to this.

Bibliography

Albrecht, William A. *Let Rocks Their Silence Break*, American Institute of Dental Medicine, 1954.

Apfelbaum, Steven I. *Nature's Second Chance: Restoring the Ecology of Stone Prairie Farm*. Boston: Beacon Press, 2009.

Arnold, Nick, and Tony De Saulles (illustrator). *Microscopic Monsters*. Horrible Science Series. London: Scholastic Children's Books, 2001.

Balfour, Lady Eve. *The Living Soil*. Reprint. Bristol, UK: The Soil Association, 2006.

Berry, Wendell. *The Unsettling of America: Culture & Agriculture*. New York: Avon Books (first Avon printing), 1978.

Bingham, Sam. *The Last Ranch: A Colorado Community and the Coming Desert*. New York: Basic Books, 1996.

Boyle, David, ed. *The Money Changers: Currency Reform from Aristotle to e-cash*. London: Earthscan, 2002.

Boyle, David. *The Little Money Book*. Bristol: Alastair Sawday/Fragile Earth, 2003.

Capra, Fritjof, *The Web of Life: A New Scientific Understanding of Living Systems*. New York: Anchor Books, 1997.

Darwin, Charles. *The Formation of Vegetable Mould, Through the Action of Worms*. London: John Murray, 1881.

Fukuoka, Masanobu. *The One-Straw Revolution*. Reprint. New York: The New York Review of Books, 2009.

Fukuoka, Masanobu. *Sowing Seeds in the Desert*. White River Junction, VT: Chelsea Green Publishing, 2012.

Gershuny, Grace, and Joe Smillie. *The Soul of Soil: A Soil-Building Guide for Master Gardeners*. Fourth Edition. White River Junction, VT: Chelsea Green Publishing, 2009.

Howard, Sir Albert. *Farming and Gardening for Health or Disease*. Reprint. Bristol, UK: The Soil Association, 2006.

Jacobs, Jane. *Cities and the Wealth of Nations*. New York: Random House, 1984.

Kellogg, Charles E. *The Soils That Support Us: An Introduction to the Study of Soils and Their Use by Men.* New York: The Macmillan Company, 1941.

Kravčík, M., J. Pokorný, J. Kohutiar, M. Kováč, and E. Tóth. *Water for the Recovery of the Climate: A New Water Paradigm,* http://www.vodnaparadigma.sk/indexen.php?web=./home/homeen.html.

Liebenberg, Louis W. *The Art of Tracking: The Origin of Science.* Cape Town: David Philip, 1990.

Leopold, Aldo. *A Sand County Almanac.* Reprint. New York: Ballantine, 1990.

Mason-Jones, David. *Should Meat Be on the Menu?* Kindle Edition/Momentum, 2012.

Montgomery, David R. *Dirt: The Erosion of Civilizations.* Berkeley: University of California Press, 2007.

Outwater, Alice. *Water: A Natural History.* New York: Basic Books, 1997.

Picton, Lionel. *Thoughts on Feeding.* Reprint. Bristol, UK: The Soil Association, 2006.

Pielke, Roger, Jr. *The Climate Fix: What Scientists and Politicians Won't Tell You About Global Warming.* New York: Basic Books, 2010.

Pollan, Michael. *The Omnivore's Dilemma: A Natural History of Four Meals.* New York: Penguin Books, 2007.

Savory, Allan, and Jody Butterfield. *Holistic Management: A New Framework for Decision Making.* Second Edition. Washington, DC: Island Press, 1998.

Schumacher, E. F. *Small Is Beautiful: Economics as if People Mattered.* New York: Harper & Row, 1975.

Shiva, Vandana. *Soil Not Oil: Environmental Justice in an Age of Climate Crisis.* Cambridge, MA: South End Press, 2008.

———. *The Violence of the Green Revolution: Third World Agriculture, Ecology and Politics.* Third Printing. London, New York, and Penang: Zed Books Ltd., 1997.

Tasch, Woody. *Inquiries into the Nature of Slow Money: Investing as if Food, Farms, and Fertility Mattered.* White River Junction, VT: Chelsea Green Publishing, 2009.

Voisin, André. *Soil, Grass and Cancer.* Reprint. Austin, TX: Acres U.S.A., 2003.

Walters, Charles. *Unforgiven: The American Economic System Sold for Debt and War.* Reprint. Austin, TX: Acres, U.S.A., 2003.

Yeomans, P. A. *The City Forest: The Keyline Plan for the Human Environment Revolution.* Sydney: Keyline Publishing Pty Ltd, 1971.

Index

INDEX

About the Author

Judith D. Schwartz is a longtime freelance writer whose work has appeared in venues from *Glamour* and *Redbook* to *The Christian Science Monitor* and *The New York Times*. She is the author of several books, including *Tell Me No Lies: How to Face the Truth and Build a Loving Marriage* (coauthored) and *The Therapist's New Clothes*. She has an MS from the Columbia Graduate School of Journalism and an MA in counseling psychology. She lives with her family in southern Vermont.

About the Foreword Author

Gretel Ehrlich is the recipient of the 2010 PEN New England Henry David Thoreau Award for Nature Writing. She lives in Hawai'i and Wyoming.

Chelsea Green Publishing is committed to preserving ancient forests and natural resources. We elected to print this title on 30-percent postconsumer recycled paper, processed chlorine-free. As a result, for this printing, we have saved:

9 Trees (40' tall and 6-8" diameter)
4 Million BTUs of Total Energy
809 Pounds of Greenhouse Gases
4,387 Gallons of Wastewater
294 Pounds of Solid Waste

Chelsea Green Publishing made this paper choice because we and our printer, Thomson-Shore, Inc., are members of the Green Press Initiative, a nonprofit program dedicated to supporting authors, publishers, and suppliers in their efforts to reduce their use of fiber obtained from endangered forests. For more information, visit: www.greenpressinitiative.org.

Environmental impact estimates were made using the Environmental Defense Paper Calculator. For more information visit: www.papercalculator.org.